Walid Kaakeh

Integrated Pest Management

Walid Kaakeh

Integrated Pest Management

A Comprehensive Guide to Sustainable and Effective Pest Control Practices

Noor Publishing

Imprint

Cover image: www.ingimage.com

Publisher:
Noor Publishing
is a trademark of
Dodo Books Indian Ocean Ltd. and OmniScriptum S.R.L publishing group

120 High Road, East Finchley, London, N2 9ED, United Kingdom
Str. Armeneasca 28/1, office 1, Chisinau MD-2012, Republic of Moldova, Europe
Managing Directors: Ieva Konstantinova, Victoria Ursu
info@omniscriptum.com

Printed at: see last page
ISBN: 978-620-8-87057-7

Integrated Pest Management

A Comprehensive Guide to Sustainable and Effective Pest Control Practices

Prof. Dr. Walid Kaakeh

International IPM Consultant

First Edition - 2025

بسم الله الرحمن الرحيم

IN THE NAME OF ALLAH

THE MOST BLESSED, THE MOST MERCIFUL

Dedication

To my mother and father,
Aisha and Abdul Ghani Kaakeh
May God bless them and make them dwell in
His spacious gardens

To my wife Fatima Khatib

To my beloved children,
Tasneem, Yasmine, Yamen, and Enas

I dedicate this book

Walid Kaakeh

Table of Contents

Introduction

Integrated Pest Management (IPM) stands as a transformative approach at the intersection of science, sustainability, and agricultural innovation. It offers a multi-dimensional framework for addressing pest challenges effectively without sacrificing environmental integrity or agricultural productivity. As the global population grows and agricultural demands intensify, the need for sustainable pest management practices has never been more critical. This guide aims to empower farmers, agricultural engineers, policymakers, researchers, and practitioners with comprehensive knowledge, innovative strategies, and practical tools to implement IPM in diverse agricultural systems. By doing so, it underscores the importance of balancing productivity with ecological responsibility.

At its essence, IPM is a dynamic and holistic system that integrates multiple pest control methods, including biological, cultural, mechanical, and chemical approaches. By emphasizing preventive practices and grounded ecological principles, IPM provides an alternative to conventional pest control methods, which often rely heavily on broad-spectrum synthetic pesticides. These traditional methods can lead to unintended consequences, such as pesticide resistance, harm to non-target species, degradation of soil and water quality, and negative impacts on human health. IPM, in contrast, offers a balanced, adaptive, and science-based strategy that prioritizes long-term sustainability. It seeks to minimize risks to ecosystems, safeguard beneficial organisms, and protect the well-being of agricultural communities.

The foundation of IPM lies in understanding the science behind pest management. This guide begins by exploring the historical evolution of pest control, tracing the shift from synthetic chemical reliance to agroecology-based solutions. It emphasizes the importance of emerging scientific advancements such as CRISPR-based gene editing, RNA interference, and artificial intelligence for pest detection and forecasting. By integrating traditional knowledge with cutting-edge technologies, IPM equips stakeholders with the tools to address modern agricultural challenges in a rapidly changing world.

Sustainable pest management relies on understanding ecological principles, such as the dynamics of pest populations, the role of biodiversity in natural pest suppression, and the resilience of agroecosystems. This guide delves into these foundational elements, offering readers insights into how healthy ecosystems can naturally regulate pest populations. Detailed chapters focus on pest identification, monitoring tools, and forecasting models, enabling farmers and engineers to make informed decisions based on accurate risk assessments. Equipped with this knowledge, practitioners can implement strategies that proactively prevent pest outbreaks rather than reacting to infestations after they occur.

To make IPM actionable, this guide presents an array of practical strategies tailored to various cropping systems and contexts. It covers preventive cultural methods like crop rotation and intercropping, biological control techniques utilizing predators and parasitoids, and mechanical tools such as exclusion nets and vacuum systems. Selective chemical applications are discussed as a last resort, with an emphasis on reduced-risk pesticides and resistance management. Case studies from around the world highlight the real-world successes of IPM implementation, offering valuable lessons and inspiring possibilities. From the integration of mating disruption techniques in California almond orchards to the deployment of push-pull technology for maize pest control in East Africa, these examples showcase how IPM can be adapted to local conditions and scaled effectively.

Despite its immense potential, the widespread adoption of IPM is not without challenges. This guide addresses critical barriers, including the need for education and awareness, the upfront costs of transitioning to sustainable practices, and resistance to change among stakeholders. It calls for concerted efforts among farmers, researchers, policymakers, non-governmental organizations, and global institutions to overcome these obstacles and foster collaboration. By scaling up IPM practices, the agricultural sector can achieve greater sustainability while contributing to global food security and environmental protection.

Looking ahead, the future of IPM lies in the convergence of science, technology, and policy. Technological innovations, such as smart pest monitoring systems powered by sensors and artificial intelligence, drone applications for precision agriculture, and the development of biodegradable and plant-based pesticides, are set to redefine pest management. As climate change and emerging pest threats continue to pose challenges, IPM offers a resilient and adaptive framework to address these uncertainties. By 2050, IPM is poised to remain a cornerstone of sustainable agriculture, fostering resilience, productivity, and environmental stewardship.

Whether you are a farmer seeking actionable guidance, an agricultural engineer advancing pest management science, or a policymaker shaping the future of sustainable agriculture, this guide provides the resources and insights to navigate the complexities of IPM. Together, we can chart a path toward agricultural practices that not only meet the demands of today but also preserve the planet for generations to come. Let us embrace the promise of IPM and work collectively toward a sustainable and impactful agricultural future.

Author
Prof. Dr. Walid Kaakeh
May 2025

Part I - Scientific Foundations of Pest Management

Chapter 1

Introduction to Integrated Pest Management (IPM)

Contents:

1.0. Introduction

Integrated Pest Management (IPM) has been and is widely used terminology; however, it is a term that very few understand adequately. Obviously, the concept of IPM consists of more than just the terminology. In an ultimate sense, the goal of any IPM program is to lower pest populations to tolerable levels. This is a very important departure from many of the more traditional approaches to pest control which were based upon the elimination or eradication of pest populations. An offshoot of this postulate is that any control measure undertaken against a pest population be economically justified. This economic justification will entail an identification of the economic injury level (EIL) of each pest within a total management system. The EIL concept is of paramount importance in the attempt to justify in monetary terms the control of a pest (Gold, 1981).

The EIL under idealized conditions is a relatively simple mathematical computation and many such models exist for many pest-crop system combinations. However, the EIL concept when reduced to practice is fraught with many pitfalls, some of which are only now being recognized. The EIL is dynamic in nature, fluctuating with a number of biotic and abiotic factors. Hence, risk assessment models must be continually run and updated. IPM as a concept has become the “buzz” or “in” phrase in Washington circles. It is being touted as the answer to all pest problems and, thus, as a panacea for many environmental problems.

Essentially, these panaceas boil down to a few principles: (1) the use of properly timed pest assessments, surveys, forecasts, traps, or "scouts" designed to warn producers of a potentially serious pest problem; (2) the use of alternative pest control technology including cultural methods, biological control, bug-proof structures, pheromones, traps, or other non-chemical methods; and (3) a reduction in pesticide inputs or a severe restriction of the use of chemical pest control technology.

1.1. What is IPM? (Definition, Concept, Core Princip les)

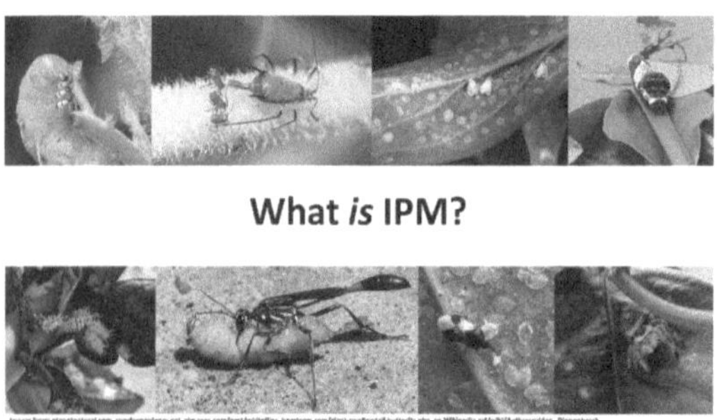

Integrated Pest Management (IPM) is a term that is widely used but little understood. Essentially, the ultimate goal of IPM programs is to reduce pest populations to tolerable levels. This may be a departure from more traditional approaches to pest control that advocated the elimination or eradication of pest populations. Another postulate of IPM is that any control measure undertaken against a pest population be economically justified. The economic injury level (EIL) needs to be identified for each pest within a total management system. There is the continuing need for both basic and applied research by interdisciplinary teams to fully implement and benefit from IPM programs. When IPM is implemented fully in approaching solutions to pest problems it does offer efficient pest control and environmental preservation, but IPM programs have not yet reached their full potential. More often than not when someone refers to IPM, what is meant is effective insect pest control. The field scouting aspects of IPM have been greatly enlarged upon and effectively promoted by . It has been the Extension Service of that has been the major in-field promoter of the concepts of IPM.

IPM is a logical extension of many previously practiced pest management approaches but has only been in widespread use since the late 1960s. Until that time, pest management consisted mostly of calendar or routine sprays applying a pesticide whether it was needed or not. Such practices became less acceptable due to the misuse and abuse of hazardous pesticides. Additionally, pest populations could increase rapidly in the absence of properly timed control measures, resulting in extensive crop damage before the pest appeared above threshold levels. Consequently, pest management has become more sophisticated, incorporating principles of population biology, ecology, math modeling, and risk assessment into decision-making programs. It involves the forecast and prevention of pest population outbreaks rather than simply the reaction to them. Thus, sound program planning will include the formulation and application of management principles intended to minimize pest problems within the growing environment.

1.2. Historical Evolution of IPM (From DDT to Agroecology)

Traditionally, insect pest management was based on the application of inorganic or organic broad-spectrum insecticides, which were toxic to both the pest and the naturally occurring beneficial organisms. Fortunately, insecticide resistance was recognized more than four decades ago, leading to the initiation of several programs for developing the concept of biological control. This endeavor was initially slow and uncoordinated.

The introduction and subsequent adoption of the first synthetic organophosphate, chlorpyrifos, in the 1970s in the Arabian Gulf countries and the Southern European and Eastern Mediterranean nations led to a new wave in agricultural intensification, with attendant economic and social benefits. But in no time at all, all these nations experienced major insect resistance problems, many of which led to significant secondary pest outbreaks. The break points for this rapid failure of chemical control were the necessary concomitant targets for economic development: to increase the area, production intensification, and crop diversification. Each exacerbated the situation, and as the problems worsened, each remedy aggravated the pest status of the very pest of concern. An awareness of the difficulties inherent to a pest, ideology, and control paradigm that was overwhelmingly chemical was ultimately recognized, leading to the call for alternative pest management approaches.

In South and Southeast Asia in particular, ocular hormones were identified as the missing link for population studies, and if the presence of this hormone could be established, the likelihood of a major outbreak was next to nil. Thus, techniques were established for detection both in the field and the laboratory, thus allowing pest forecasts to be made some 4-6 weeks in advance of an impending break of pest numbers. The reaction occurred a week in advance, allowing control measures to be preemptively applied. Yet again, the controls were predominantly limited to chemical solutions, relying again on naturally derived compounds. In addition, to say that the enormously generous budgets of the UNON project were utilized inefficiently would be an understatement. But as a measurable and enviable success story, the last carpet-level treatment of organophosphate was successfully implemented in 1994 in Iraq (Pretty & Pervez Bharucha, 2015).

1.3. Economic Threshold and Economic Injury Level in Pest Management

The concepts of Economic Threshold (ET) and Economic Injury Level (EIL) are fundamental to Integrated Pest Management (IPM) systems, providing scientific guidelines for when pest control measures become economically justified. These concepts help balance agricultural productivity with environmental sustainability by preventing unnecessary pesticide applications while protecting crops from significant economic damage.

Core Definitions

- Economic Injury Level (EIL): The smallest number of insects (or amount of injury) that will cause yield losses equal to the insect management costs. It represents the "break-even point" where pest damage justifies control expenses.
- Economic Threshold (ET): Also called the "action threshold," this is the pest density at which management action should be taken to prevent an increasing pest population from reaching the EIL. The ET is always set below the EIL to allow time for control measures to take effect.
- Key Differences Between ET and EIL:

1. Nature**:** The EIL is a theoretical value representing the point where economic damage begins, while the ET is a practical, operational rule for decision-making.

2. Position: The ET is always set below the EIL to provide a "lead time" before economic damage occurs. For soybean aphids, for example, the ET is 250 aphids/plant while the EIL is 674 aphids/plant.
3. Function: The EIL serves as a benchmark for economic damage, while the ET triggers management actions.

Calculation and Components

The EIL can be calculated using the formula:

EIL = C/VIDK Where:

- C = Cost of control per production unit
- V = Market value per production unit
- I = Injury units per pest
- D = Damage per injury unit
- K = Proportional reduction in pest attack from control.

For example, in apples:

- If control costs $20/tree (C)
- Apples sell for $1.50/lb (V)
- Each beetle damages 5 apples (I)
- Each damaged apple loses 0.5 lbs (D)
- Insecticide provides 90% control (K=0.9)

The EIL would be 5.93 beetles per tree.

Types of Economic Thresholds

1. Fixed ETs: Most common type, set at a fixed percentage of the EIL (e.g., 50-80%).
2. Descriptive ETs: Use population growth models to predict when injury will reach the EIL.
3. Dichotomous ETs: Use statistical procedures to classify populations as economic or non-economic.
4. Nominal Thresholds: Based on practitioner experience rather than calculated EILs.

Practical Applications

ETs vary significantly by crop and pest:

- Soybeans: Bean leaf beetle ET at 80% of EIL.
- Sweet Corn: Treat when 25-30% of whorl-stage plants have European corn borer larvae.
- Apples: European red mite thresholds range from 2.5 mites/leaf early season to 7.5 mites/leaf later.

- Potatoes: Colorado potato beetle ET at 20% defoliation with specific larval counts.

The ET/EIL framework remains central to modern IPM, helping growers make economically and environmentally sound pest management decisions while maintaining crop productivity and quality.

1.4. Why IPM? (Economic, Environmental & Health Benefits)

Since the introduction of IPM, numerous studies have demonstrated that IPM can provide long-term economic, health, and environmental benefits to farmers and society under specific conditions (Zhang, 1970). Farmers, policymakers, researchers, and extension programmers around the globe have been actively promoting, implementing, and studying IPM, with the aim of achieving MIPM goals. MIPM is not static; it is a constantly evolving procedure encompassing interactions among important stakeholders, farmers, government, industry, and researchers. Hence, it needs to be continually monitored, controlled, and improved.

Any pest management action is not within an isolated natural ecosystem or agroecosystem. It is also a social economical action. The economic aspect of pest management (pests or pest management methods) is one of the very important factors influencing the outcome of the pest management. After each possible action, pest management outcomes (economic, environmental, health or other societal impacts) should be assessed and the feedback would be used to adjust pest management strategies. All economic costs and benefits should be included in the assessment. Economic costs including production costs, agricultural product price impacts, switching costs, and risks cost. Different management practices with variations of both economic costs and benefits should be included to minimize the overall economic effect of pest management before the time threshold.

1.5. IPM vs. Conventional Pest Control (Pros & Cons)

Integrated pest management is a multi-faceted approach to satisfying humans' desire for food, fiber, and other biological products while minimizing impacts on health, the environment, non-target organisms, and future crop production. In a society as a whole, time, effort, and environmental quality must be weighed against output in food and fibers. Broad and short profit calculations may guide short-term decisions, but the question of the long-term sustainability of established production methods via IPM is largely beyond the scope of the agricultural research and extension communities. It is clear that the effectiveness and acceptability of any pest management system is a matter of these broader (and more subjective and less easily quantified) social, political, and economic considerations (Pretty & Pervez Bharucha, 2015).

Among these considerations are trade-offs between environmental quality and agricultural sustainability. Where there is an adequate supply of unprocessed land, water, labor, and other production resources, the long-run sustainability of crop and animal production systems has tended to be taken for granted. However, with increased land degradation, water scarcity, labor shortages, and likely climate change effects, this is no longer appropriate. Agricultural intensification and structural change options for maximizing long-term output minimizing trade-offs, and meeting societal demands for non-agricultural land and environmental quality are therefore now urgent global issues. The impacts of IPM on the long-run sustainability of agricultural systems are difficult to assess, and hence the balance of various costs and benefits neither easily quantifiable nor universally accepted.

The system approaches involved, simplicity, as well as costs and benefits in terms of use of 'cheap' resources, equity regional and intra-household allocation, cultural factors, and perceived changes in agricultural prospects all affect and help explain differences in farmer adoption of IPM. As in other countries, almost all farms in most Indonesian and Philippine municipalities, regions, and islands might adopt because of 'contagion' effects in information flow and technology development and dissemination. However, where squatter settlements provide sudden dramatic changes in farm populations that agricultural research and extension systems cannot match, as in Sulawesi, the situation is very different.

1.6. IPM Opportunities and Challenges

Integrated Pest Management (IPM), like climate change, sustainability and food security, is well-established as a global priority but much remains to be done. IPM is expected to control pests, increase agricultural productivity, safeguard the environment and help improve rural livelihoods (York State Integrated Pest Management Program, 2016). IPM offers those involved in global agricultural development a distinct opportunity to help promote sustainable agriculture and food security. The goal is to elaborate on the need for IPM to control weeds, diseases and invertebrate, vertebrate, and plant pests that damage crops, livestock and food stores. Much of the developing world's agricultural landscape now comprises smallholdings of less than one family fungi, insects and weedy plants (Parsa et al., 2014). Controlling pests and pathogens, and managing competitive weeds, is difficult with limited resources but, if ignored, these pressures prevent attempts to grow more food for an increasing global population. General win-win opportunities arise from increased demand for food and other commodities, together with international development funding, and for increased and more effective engagement with the private sector to increase production in developing countries and to promote fair trade.

Many opportunities exist within current international initiatives for an increased effort to promote IPM. Continued pest pressure from climate change influences will accelerate the need to tackle this threat. IPM innovations that have an overall greenhouse gas (GHG) emissions reduction potential but that nevertheless correspond to GHG burdens at the level of individual practices may need special consideration, aimed at promoting farmer-education and ecological pest control, horticultural export APPS IPM, non-regulated pesticide registration, capacity-building in participatory NARS/NGOs and promotion value-chains of farmer cooperatives for bio-R&D, focusing on biocontrol and botanical-derived pesticides, especially for tropical regions. Promoting IPM presents an exciting opportunity for all those concerned with development in the 21st century.

1.7. Environmental and Economic Impacts of Traditional Pest Control Methods

The consequences of pesticides on environmental and economic sustainability are particularly compelling. Pesticides are toxic to all biota and so they have the potential to be detrimental to ecosystem structure and functions. Habitat modification resulting from modern high-input agronomy has already resulted in decline of biodiversity in general, and declines of top predators and other natural enemies of pests are likely exacerbating pest problems. Institutional and financial failure to support organic and IPM will lead to increased costs for lower yields, increased reliance on pesticides, pest resistance and resurgence, biodiversity loss, etc. Consequences will be real, and potentially devastating, unless both government and non-governmental organizations (NGOs) act aggressively to redress these failures (Pretty & Pervez Bharucha, 2015).

To be economically sustainable agriculture must not only meet the primary economic needs of food, fiber, timber, etc. but also take into account secondary market and non-market economic services provided by other elements of "natural capital." This includes pollination of crop and wild plants, carbon fixation and storage, degradation and ultimately storage of toxic biota, biota control of pest, pathogen and weed irrigation, and soakage of excess water. Traditional agriculture depends on these processes of nature, which are cheap and often free. Even under ideal closed reference conditions with little or no anthropogenic change they have tended to change over time. Complex agriculture systems are fashioned which maximize the evolutionary stability of the diverse components and their processes. This results in relative stasis. The stability of these systems can be driven into open reference condition by modified plant forms or the introduction of invasive biota, which disrupt the processes controlling pests, weed and ecosystem change.

The shift to simpler proprietary agriculture systems amenable to chemical fixes first led to yield increases but all higher yield genotypes depend on lower input agriculture stocks for their selective improvement and the inputs are increasing in price or supply is diminishing. This means that in order to remain competitive growers have little option but to continue appropriating nature. So the pest resurgence cycle drives agriculture toward a new equilibrium level with increased numbers of application of more powerful cheaper pesticides. However this is not without costs. The pervasive use of these chemicals leads to broad scale declines in the numbers of predators and parasites pernicious to pests and increased prevalences of pest species and higher population densities. Original pests, new invasive species, secondary pest outbreaks, increased dollar losses/crop.

1.8. Role of Farmers and Agricultural Engineers in Sustainable Practices

A multidisciplinary approach to pest management which integrates technologies and knowledge developed by different disciplines offers hope for capacity building in the quest for pest control by the African farmer. Successful and sustained application of this technique would provide the basis for the development of pest control strategies and technologies which are regionally adaptable to different landscapes and socio-economic constraints. Control interventions which take advantage of the natural biological control mechanisms and selectively apply appropriate technologies would enable the African farmer to operate close to maximum cost-benefit efficiency. This would drastically reduce waste in the pest management system and result in large savings. These savings would ameliorate poverty, promote food security, and also advance the development of pertinent technologies to move agricultural production forward in the emerging global economy of information and cybernetics (Okigbo, 2018). Integrated Pest Management (IPM) is a combination of practices that control pests effectively and economically while minimizing risks to people and the environment. It is an integral part of sustainable agriculture. The primary keyword of IPM is integrated. IPM includes biological control and environmental management using antagonistic animals, natural enemies, resistant plants, cultivation techniques, and ecological principles. Conventional pest control techniques are based largely on analysis and prediction, while integrated management methods take into account both the pest and the pest control technology. The emphasis is shifting from product development to ensuring sustainable agricultural productivity by resource preservation. This type of cost reflects environmental aspects of sustainability which is a critical element of IPM. The bioeconomic principles provide decision guidelines that aid farmers to optimize returns on pest management activities while paying due attention to social effects and maintenance of environmental quality (Quarcoo et al., 2017). Agronomic and other crop production practices usually provide benefits beyond the immediately apparent ones to crop producers. The objective of recommended agronomic or cultural practices is to make plants more healthy and productive. This requires that specific practices be adapted to soil-water-plant-climate conditions, and organizational and economic considerations. Each agronomic practice can be achieved using a range of specific methods. For example, plants can be made more healthy by improved soil fertility management through additions of organic matter, fertilizers, and other growth-promoting substances. Techniques that improve soil physical properties, alter soil chemical properties, or increase soil biological activity or any combination of these can accomplish this objective. Topography, other natural resource availability, labor availability, land tenure, and farmer knowledge and information have to be considered when identifying methods for particular circumstances. Thus, managers of individual farm enterprises must opt for combinations of methods that make economic sense for their scale instead of solely relying on methods used by successful large-scale farms.

1.9. Quick Self-Assessment – Is Your Farm Using IPM?

Take time every now and then to remember that you're the one in charge. If you encounter major problems that don't seem amenable to even partial solutions, call your pure questions instead of just greater troubles. Remember that what is done is done, but to imagine undoing it is impossible. Increasingly push your thoughts into the future and consider possible good directions and situations with sufficient clarity to see what constellation of things, actions, and events need to remain constant, bad, or to be avoided. Practice, if only occasionally, during periods of solitude and quiet time, a rigid sort of attention discipline, an ability to focus upon a singular thought even against determined distraction attempts (Pretty & Pervez Bharucha, 2015). Think on it long enough until it seems obvious, reflexive, even.

Review these instructions each day for a week. Modify them or adapt new ones as seem necessary or more appropriate. For better or for much worse, these will inevitably reshape your life, manners, perspective, evaluations, beliefs. Embrace the idea of doing so. This raises the larger question: what do you want that reshaping to look like? And what is there to do about it? This question is unwieldy and impossible in its entirety. Tackle a smaller variant such as what would the future life that you would like to live be exactly this evening. To generate the plausible best answer and even sketch it out, spend time thinking about it (Stetkiewicz et al., 2018). Make 100 pots an hour or 1 pot per day, become an architect or a programmer, pull whole weeks of assembly-line shifts or continue doing research indefinitely, settle down in Kansas City, travel the world, rip it up, initiate a family, spend evenings staring out the window, be angry, be sad, be still – this is powerfully possible, and will unfold more in time and with practice.

Chapter 2
Ecological Principles in Integrated Pest Management

Contents:

2.0. Introduction

Pest organisms can be managed effectively by knowing pest population-limiting factors and manipulating the targeted factors. Organization of pest control mechanisms in a framework containing principles and processes of population dynamics, community organization and functioning, ecosystem organization and functioning, and evolution in the long term can be established as an integral part of pest management. An understanding of pest population dynamics helps the pest manager in making predictions regarding pest abundance on the crop, and also provides the basis for furthering research to fill the knowledge gaps regarding beneficial species and processes involved in pest suppression. In the pest management and agro-ecosystem health studies, it is more advantageous to devise approaches in pest population dynamics, at individual and community levels, for pest management on the farm, instead of concurring with attempts for understanding pest damage patterns and crop loss assessment. Population dynamics of a pest can be studied in relation to one or more of its regulating factors. A plethora of abiotic and biotic pest population-limiting factors exist. In pests of agricultural crops, the economic damage threshold depends not only on the economic status of the respective crop but also on biological and subjective factors.

Knowledge of pest population limiting factors paves the way for devising effective pest management methods. Depredations by natural enemies such as predators, parasitoids, and pathogens keep the target pest species at very low and tolerable levels. Other final effectors of pest population decline, such as habitat interspersing, alien species introduction, agricultural practices, non-host trap cropping, semiochemicals, and reproductive controls, are less effective since the crop is monocultural. In the absence of the final effectors, field pests can build up at a released rate. The understanding of such processes helps in devising pest management systems which are ecologically sound. Sea buckthorn plants harm the geometry of coffee fruit borer predation on its pupal stage and plausibly inhibit the parasitism on pupae processes for this pest. Population modulating effect of temperature on pest damage on the soy crop has also been documented. Integration of pest management research in the crop defence system, ecology and economics is the need of the hour for ensuring food safety in a sustainable manner.

2.1. The Role of Biodiversity in Natural Pest Suppression

Biodiversity refers to the variety of life found in a particular habitat, including the different species of plants, animals, microorganisms, and the genetic variation within each species (Quinto Cánovas et al., 2010). This richness supports ecological processes, such as energy flow and nutrient cycling that contribute to the maintenance of conditions for life on the planet. Biodiversity compartments currently considered are genetic, species, and ecosystem diversity. At the genetic level, the number of alleles lead to a degree of heterozygosity affecting fitness; species diversity is defined by the number of species and their relative abundance; ecosystem biodiversity refers to ecosystems recognized in terms of vegetation or geological typology, the degree of conservation of soil and water, and their species richness (Francis et al., 2020). Biodiversity has been defined as the totality of genes, species, and ecosystems in a particular location. International conventions have promoted some biodiversity strategies to protect this resource, and there is increasing awareness of the need to protect biodiversity at a large scale. Currently, many national and regional protocols are being put into action. Agricultural intensification, although improving food production capacity, has caused the overall reduction of biodiversity in agricultural ecosystems. In integrated pest management (IPM) safety, effectiveness and sustainability of methods employed are based on the principle of conserving biodiversity within the agricultural landscape. Biodiversity at different levels usually is of help in pest management, since its increase normally correlates with a decrease in pest population size. Strategies currently designed to promote natural pest suppression by biodiversity conservation are of predictive nature and usually require a profound knowledge of the agroecosystem. This know-why is currently lacking in many crops and agroecosystem configurations. However, effective decisions can usually be made based on field observations and a landscape perspective. When possible, it is advisable to model the association patterns between pests and their natural enemies for their effective predictions in untested crops or landscapes.

2.2. Agroecosystem Analysis

Agricultural systems are complex entities that produce food and other desired agricultural products and social and ecological services. Understanding the complete agronomic production system in space and time from multiple points of view and disciplines is required to obtain more insight into the complexity of interactions occurring in natural and managed agriculture. Negative side effects of the production systems are externalities not paying even if private costs are equal to social costs. Nevertheless, the natural performance of agroecosystems provides a wealth of options to innovate agro-production systems that take these externalities efficiently into account. Agroecosystems respond to market-driven or technology-driven changes with unintended other changes, far outside the original intention. These shifts in the system require better predictions on the consequences of interventions or alternative paths in a multidisciplinary framework outside the existing disciplinary paradigms.

Systems approaches are well-suited to provide these insights, to analyze and optimize existing situations, and to design new sustainable agroecosystems. A system comprises a set of interdependent components. The interaction or feedback structures among the components are commensurately to the connections that determine how system outputs (and the other varying joint features) are affected by changes in the inputs . The dynamics of a system may either be studied with respect to equilibrium states (situations in which the state variables remain constant over time) or transient states. The components may be described in terms of stocks, states or continuous variables, and flows or discrete events. Hence, systems can be either continuous or discrete, and either deterministic or stochastic. Systems may also be classified in a hierarchy according to their overall structures. Systems with a hierarchical or decomposable structure may be analyzed in sub-systems with the same type of variables, separating internal and external dynamics . Also, systems may be either adaptive (able to modify their interaction structures and variable parameters) or directly used, non-adaptive, common engineering implements. Agroecosystem interactions and feedback structures often occur over long distances and time spans, and engineering developments at and for a local scale can have unexpected consequences at a regional or national scale and also far more actions downwards generations.

2.2.1. Soil Health

Soil health is a fundamental element of natural ecosystems. Cropland and pasture expand the biotic component of soil health. These changes are mitigated by disturbance and discontinuities in the continuation of soil training. This chapter finds a connection between core soil health concepts. A sustainable soil health management system was defined. Soil health is an indicator of soil quality management incorporating biological, physical, and chemical processes and properties. Soils with continual soil training are considered more sustainable than those severely disturbed by anthropogenic effects. These sustainable systems are relatively composted, structured aggregates, with greater microbial diversity and abundance, pore sizes, and connectivity. Past and present analogues of these systems were detailed (W. Doran & R. Zeiss, 2000).

Soil is the life support system on Earth. The soil biota is a large and largely unknown component of that system. The biota in the upper 15 cm of soil is incompletely inventoried, yet a 1990 estimate is 1024 cells or organisms and 1030 viruses. To sustain life, soil must "hold water," convert water to a solution that contains bioavailable neutral energy and electron sources, and retain organic matter and microbial biomass which, when let loose, are biosynthesized into protean life forms.

The soil is where weathered geo-material; biogenic allogenic product were transported and unreleased; organic material sedimented and decomposed; catabolic byproduct diffused as CO2, O2, etc; and eukaryote-diagenic life is. The soil is organized in intricate patterns and nested hierarchies. There are large regional aggregates and structures and local micro-aggregates and structures; soil is very porous yet impermeable. Primary mega aggregates are mainly mineral, clay, silt and quartz sand. Several sand kinds, silt and medium sands, felts, are bio-friendly aggregates. The bio-friendly aggregates retain porous and surface, and metals such as Ca, Fe, Mg and medi-biolites in or on their surface. Condensed and solid bio-films, biological subject, biogenic aggregates are all nano aggregates.

2.2.2. Plant Diversity

In general, the implication of biodiversity in IPM is better understood by considering the following levels of biological organization in response to agriculture: Genetic diversity, species diversity, functional diversity, and community diversity (Schuldt et al., 2019). Plant diversity decreases with agriculture because management of agricultural lands focuses on a single or a few crop species which are cultivated as monoculture or with a few other crop species. Supported by chemical control, the abundance of crop plants and crop insect pests increase at the same time. This increases the trophic level of agriculture from two to three and makes the ecosystem unstable by the influx in ecosystem-food web complexity.

Though diversity can decrease sustainability, agroecological planning of agricultural systems offers management opportunities to increase relatively autonomous crop and pest diversity by including companion or ancillary flower plants (Okigbo, 2018). Diversification mediated by management is possible at a lower intensity of agriculture. Studies have focused on apid bees that provide pollination, but it can be generalized to all trophic levels of primary producers and plant consumers. Agroecosystems provide diversity of resource at many biological levels. It is therefore not surprising that management-induced diversity at one biological level generates much diversity at higher biological levels and make the agriculture more sustainable against pest emergence and outbreaks. On the other hand, it is understood that the provision and population dynamics of alternative food resource is necessary for pest control or biodiversity conservation.

Ecosystem-type loss in agriculture can occur simultaneously with geographical expansion of agriculture to keep biomass productivity. However, this mechanism cannot last long. Magnitude of land-use change will saturate and further loss in bare agricultural land will reduce the provided mosaic by agricultural operations. Target pests can also become resistant to exotic alien chemicals in pest control. A self-stabilizing agroecosynthesis to reverse the canker of diversities from past loss was suggested. Between 10-15 years after the pesticide and herbicides was phased out, the biodiversity attracted by wildflower planting showed self-repairing ability to maintain plant growth by self-managed biotic interactions at a lower resource expenditure.

2.2.3. Pest Resilience

Agroecosystems are complex ecosystems because they are affected by biogeoclimatic factors, soil types, cropping systems, rotations, and sequencing of management practices (Okigbo, 2018). Any agricultural or pest management practice can alter the diversity, niche and functional group structures of specific assemblages. Environmental circumstances, by definition, refer to all relevant factors except those pertaining to the organisms themselves. These include either specific or non-specific factors. The success of pest sampling plans depends heavily on the choices of sampling plot size, shape, configuration, and, when relevant, trapping density in the plot. In order to be better at directing allocation of sampling effort toward those areas where control measures are likely to be most needed, capturing the greatest possible diversity of conditions in the surrounding arena is desirable. However, as plots become smaller, capturing the full heterogeneity of the controlled variable becomes increasingly compromised. Pest management is better served by approaches that allow for event-specific assessments of pest population density, complemented by empirical models that estimate population (reproductive) failure probabilities in unplotted areas, given observed conditions in any sampled parts. Such approaches reveal broad spatial heterogeneity in pest population conditions—that is, aggregation of pest density at scales greater than that captured by fields, which means that further increases in efficiency could be gained from planning for larger sampling plots.

Resistance to insecticides is an example of responses of a pest population to human interventions. It was recently shown that when a first generation of response-locking insects are exposed even for a short time to pyrethroids, the involvement of even non-target cation channels in resistance can become genetically fixed (because their genetically wired responses to insecticides are the same). Insecticides cannot completely annihilate a parent generation of response-locking insects, which, however, still manifest a marked reduction in density (through both birth-death and contribution dispersals at time n, in which remaining insects were not wholly exterminated). Regardless of any changes in the properties of the underlying landscape, the overall pest population performance is nevertheless impaired. This deprivation of pests out of high suitability areas results in the appearance of extended unsuitable zones that were previously of high suitabilities. These in turn raise the likelihood barriers and hinder the efforts of pest resurgence. Control of any pest population with a response-locking insecticide can thus impose a greater likelihood of long-lasting pest manageability.

2.3. How Diversified Farming Reduced Pest Outbreaks

The greatest problem of modern agricultural practices, mostly nitrogen fertilizers and synthetic pesticides was made to boost productivity ((Anton) Muhibuddin et al., 2011). The consequence of such practices contributes to negative externality. Although efficacy in controlling pest was achieved, the ecosystem unpredictable aberration may occur in terms of pest emergences or outbreaks. More than 20 years ago, an intensive pest control practiced in African vultures crop fields results in fall armyworm outbreak (Okigbo, 2018). Local people believed that absurdly, by using synthetic pesticides, it somehow caused the pest outbreak. This phenomenal anecdote demonstrates that, although already aware of ecological principles, pests emerged and outbreaked. The spatio-temporal pest outbreaks and emergence have been studied by scientist, for examples in cultured fish, tapioca, and several crops. In fiction physic, explained that pest became aggregated scarce at very early years. Density depended birth was explained by “triangles”. “Bump” offered an arthropod burrowed from the compost burrow into sandwiches. elucidated pests population explosion in classic text. These previous works revealed that, pest population growth and control were best explained by interactions among density depended historic deterministic dynamics, “shocks”, and short autocorrelation random noise. However, general public could not grasp those equations, because unlikely the simple cartoon-like representation of the “banking perfect cape graphs with constant discount rate”. Ecological principles in pest control as elucidated above seemed complex and unintelligible. Moreover such quantities, equations, and modeling output were misunderstood but seen as unequivocal “answers”, or worst, were abusively misused, such as incorporates into pest battle loss indexing schemes. It stands to reason then that, similar to cartographers' and astronomers' tasks to prepare navigational maps, whereby, "complexity was tamed in a way that made sense" to the general public, pest control scientists' objectives should also be in the similar spirit.

2.4. Ecological Interactions in Agroecosystems

An agroecosystem includes both biotic and abiotic individuals which interact with each other in various ways. A pest can be defined as any organism of animal or plant origin, namely, insect, bird, nematode, weed, etc., that inflicts economic injury to agricultural produce or food materials in storage. Crops cultivated in monoculture may encounter an invasion of herbivorous pests and diseases attacking the crop species, which act as the food and living resources of the pest. A dramatic upsurge of the pest population would generate a sudden application of environmental chemicals to reduce the pest numbers, commonly referred to as chemical pesticides. Chemical control methods rely on highly efficient chemicals to kill the invading target pest. Because the pest is just one individual of the household ecosystem as a candidate for resilience, selective pressures for resistance are anticipated against application of chemical pesticides.

The successive borrowing of chemical pesticide protection invites unexpected counteractions adopted by the environmental ingenuity of the invading pest. During the one-hundred-year history of chemical-planned pest management, the pest knowledge gap has been filled in with new pest species. Therefore, chemical control cannot provide a sustainable solution for pest problems. The ideal pesticide changes pest locomotion, chewing, sucking, subsequent internal metabolism of pesticide component, or/and physiology of metabolism to reduce to an environment-friendly material before all metabolic target sites are blocked. Control systems relying on such highly engineered pesticides would be called safer pest management systems. The stability of the ecosystem is realized when the biota are assembled as a predator-prey system. Each species has its time constant to decrease or increase their number because of eco-evolutionary responsibility in the species that cohabits with a pest.

The objective of pest management is on the conservation properties, allotted with roles to sustainability recognized in coevolutionary stability jointly formulated by equations: (1) fast growing species should be protected with superior tools, and (2) real time management has to be done on today's data on artillery and targets. Conservation Properties do not dismiss the use of chemicals but advocate a sculptural manner of the use, in regard to the format of pest onslaught, starred positions on the map of imperfect knowledge of pest population density, coots/innocent harmless individuals etc. the goal of pest management is given as "attainable as sustainability preserving properties to be recognized".

2.4.1. Predator-Prey Dynamics

Food webs are a common source of structure and complexity in natural systems. Accurately representing the extent and structure of complex food webs with classical differential equations is intractable. Previous studies have demonstrated that even modelling simple general food web models with only the direct effects between species can exhibit complex dynamics. One such model consists of a system of two species differing by their body size. All predator-prey interactions are described by functional responses, which relate the prey consumption rate by predators to the prey population density. When functional responses are represented by Holling type I responses, it has been shown that by varying the body size ratios of these two species complex dynamics can arise. This results occur for all types of functional responses, suggesting that the body size ratios play an important role in generating complexity (Therese Langeland Næss, 2016). Figure 1: Complexity versus the predator-prey size ratio. The number of complex dynamics for the three types of functional responses are shown. The bifurcation parameter is the body-size ratio of the predator to its prey log 10 (αp/αh) are shown. A codimension-two Takens-Bogdanov bifurcation can also be observed (left). The same fashion from complexity is exhibited. For Holling type III responses similar bifurcations as in type I and II transition into chaos disappearance is observed (right). Numerical simulations are used to study two species food webs with one predator and one prey. The prey is dragged from the fourth quadrant into the first quadrant. A bifurcation diagram describing how the dynamics of this simple model change by introducing feedbacks from shared food sources is constructed. As a parameter increases large ranges of complex dynamics are observed. As the predator acquires larger edge lengths to its food web the dynamics cease to be periodic and settling into chaotic behaviour. There are also ranges where chaotic solutions go extinct due to periodic windows or homoclinic bifurcations (Ananth V. S. & Vamsi D. K. K., 2021).

2.4.2. Competition Among Species

In ecology, competition refers to the struggle among living organisms to meet their requirements for life and growth. Plants compete for light, moisture, and minerals in the soil, while microorganisms may compete for nutrients in all habitats. In ecosystems, plants and animals compete for sunlight, space, and food. In a mixed crop, some crop species may not yield as much as those in monoculture since they compete for sunlight and moisture. Nevertheless, variety cropping practices can maintain a threshold level of pest populations, resulting in reduced severity. Results of both key stroked and stability, accompanied by comparative abundance, indicate that the resultant pest diversity arising from integrated management usually will not be significant. Nonetheless, if a conservative strategy is permitted to alter, often a considerable response against plant growth occurs.

Grasshoppers have a high probability of becoming invasive to new regions and locusts vary greatly in abundance and distribution as well as in their social behaviour and the extent of gregariousness. There is increasing focus on biological control of grasshoppers, especially in the western United States. However, there is currently very little research directly directed to the biological control of locusts. Most of the studies on biological control of Acrididae tend to be concerned only with phytosociological implications, conducted in misunderstandings of complex systems and co-evolution of pests-predators. Some clearer possibilities or perspectives have been stated. However, the field is still wide open for finding new species with much wider ecological effects on the target pest or phloem-feeders attacking wheat crops.

Chapter 3
Pest Identification & Monitoring

Contents:

3.0. Introduction

As populations of pest insects grow and nontarget effects of indiscriminant chemical use accumulate, farmers are looking for alternatives to conventional pest management. Integrated pest management (IPM) strategies have proven valuable in adapting to tighter pest management budgets, changing markets, and more sophisticated pest control measures (Pope & DeWitt, 2005). Location, species, and stage-specific IPM programs rely on accurate knowledge of the populations of pests and beneficial species, factors which affect their infestations, and the timing of developmental stages of the crop and each pest and beneficial. Knowledge of the local presence of pest populations is particularly critical to success because it permits focusing attention on the potentially damaging species, times, and crops. Knowledge of the factors affecting when, where, and how badly or intensely a pest and beneficial will attack improves the chance for successful pest control. Finally, knowledge of the timing of the senses, events, and actions of crop and pest life cycles may permit prevention of pest action or often reduce the amount of crop loss, cost, and damage by maximizing the chances of timely pest and beneficial action. Monitoring is needed to acquire these types of information and to help farmers anticipate pests and their future activities, similar to how weather agencies monitor the weather to predict it. Pest management forecasting is difficult but is essential for successful pest control. However, these types of activities have generally received little attention, as they do not usually have visible or immediate rewards and are often viewed as unscientific. However, they are crucial. Some major players in pest management forecasting in Iowa offer a range of information and services on local pest abundance, emergence patterns, selection pressures, plant development stages, and other pertinent issues. Many of these services are provided at no charge with very limited requirements for cooperation.

3.1. Key Pest Species (Insects, Weeds, Pathogens, Nematodes)

Agricultural pests are diverse organisms that feed or live upon agricultural commodities, reducing their availability and lowering their agricultural and economic value. Agricultural pests include insects, nematodes, diseases (pathogens), and weeds. Some key pest species in Midwestern crop production will be reviewed. A whole-field approach to pest management acknowledging that crop production is conducted in an ecological arena by interconnected producers of food will also be suggested.

Most pest management action thresholds are based on research with single pest species. Multiple pest management situations, however, are much more common than single pest situations (Pope & DeWitt, 2005). Weed management in corn hybrid beans is a good example of such multiple pest plant systems. Chemical sedge control studies in corn have been tightly controlled by establishing pure stands of corn and insects or diseases. Corn borer management has been studied in corn alone.

A good example of a complex pest situation is the following. Insects feeding on a weed nitrogenous plant infect two different crops. This injury influences weed and cropping system productivity, plant phenotype, and reproductive biology. With pests feeding on several different crops or weeds, two ecology and management strategies will be indicated. The first is the simple, classic competitive exclusion rationale in which the fewer the pest types, the better the management options. Intercropping may maintain or improve yield stability or increases in mixed-plant agriculture.

The second is the more complex pest diversity inoculation paradigm in which greater diversity on pests may make agricultural systems more desirable. Multiple pest systems with non-overlapping risks were shown to enhance plant productivity; however, it was very difficult to measure pest dynamics and inter-pest densities in multiple species systems. Nevertheless, because of the rigidity and simplicity of single pest pest management strategies, a great deal of effort has been put into understanding the ecology of insect or pathogen "super" pest conservation and detector logic intervention (D. Coble, 1989). It is not clear how many of these approaches have been successfully developed and applied to pest management on the ground rather than on paper.

Conventional cultural pest management is based on firmness and rigidity of strengths and weaknesses of differences in practice. It is a "mosaic" approach to agricultural pest management in which cropping practices and propagules are non-uniform and dispersed. Inherent inter-field variability of soil, climate, and geology, and farmer variability in systems, weed reservoirs, and interventions provides opportunities for diverse pest ecology and dynamics in management practices.

3.2. Beneficial Organisms (Predators, Parasitoids, Pollinators)

The high inclusion of natural enemy conservation methods in pest management plans suggests widespread interest in habitat manipulation to promote beneficial insects. However, we must learn more about the effectiveness of specific habitat structures intended for this goal to maximize our understanding when implementing practical solutions. Given the growing body of evidence that emphasizes the need for natural controls for pests, attention should be given to the sustainable inclusion of other farms and land types in management strategies to conserve beneficial insects (Rourke, 2017). The most commonly studied habitats for their potential benefit to natural enemies of agricultural pests were filter and herbaceous strips, hedgerows of woody perennials, and constructed, semi-natural habitats. Filter strips were least studied while hedgerows were the most studied natural enemy conservation habitat type. Each includes a significant evidence base that supports the potential benefits of conserving these habitats for improving biological control and pollination services as well as increasing pest suppression on consumed crops. Further studies evaluating how other habitats can be used for conservation purposes are warranted and this review highlights the importance of such studies. Sustainable management practices that conserve natural enemies include planting non-crop food sources to promote longevity and fecundity of natural enemies, avoiding pest control practices detrimental to natural enemies, and sustainable tillage management. Despite this wide range of options, few studies directly examined the effectiveness of these practices for beneficial insects. The high diversity of practices for managing detrimental pests includes the need for farm-specific research and implementation of recommendations. Furthermore, each of these sustainable management practices does not require land modifications and can be implemented with little to no aid.

3.3. Monitoring Tools (Traps, Scouting, Remote Sensing, AI-Based Detection)

Monitoring is absolutely essential for developing a comprehensive understanding of pest occurrence and density in various time frames and specific locations. The knowledge of pest occurrence must be meticulously complemented with thorough damage assessments if effective pest management strategies are to be undertaken. A diverse variety of monitoring tools for a single pest management system may be selected according to the complexity of the pest problem being addressed, as well as the frequency and intensity of pest monitoring. Monitoring methods can be broadly categorized into two main types: in situ (which includes techniques such as field scouting and trapping) and remotely sensed techniques, each with their own advantages and disadvantages. In situ techniques are often labor-intensive and/or time-consuming, but they typically require less advanced technology compared to the sophisticated remotely sensed techniques. To enhance the accessibility and efficiency of in situ tools, many companies have been proactive in manufacturing low-cost and simple designs that facilitate deployment in numerous regions around the globe, consequently democratizing access to effective pest monitoring technology. Furthermore, in the past two decades, a plethora of commercial technologies and services have emerged in the market, which are based on innovative smart traps that are equipped with basic cameras and weather stations. These advanced smart traps allow for effective pest-problem mapping alongside temperature, humidity, wind, and other environmental factors in near-real-time. The evolution of these monitoring techniques, particularly those integrated with the use of unmanned aerial vehicles (UAVs) in the vicinity of urban areas, has been highly favored by pest-control companies due to the significant advantage they provide in terms of capability to efficiently view and assess large areas with relative ease. (Batz et al., 2023)

Traps play a crucial role, particularly when it comes to monitoring pests in agricultural areas that are not only economically important but also highly intensive. These traps can be classified into two main categories: active traps, which intercept pests, and passive traps, which attract pests using lures. For any trap—whether active or passive—to be effective, it is essential that it undergoes thorough testing for the specific species being sampled. Additionally, the sampling techniques employed need to be carefully assessed under various conditions to ensure reliability and accuracy. When it comes to the interpretation of the capture data, it is important to keep in mind that a number of assumptions must be checked and validated to derive meaningful insights from the gathered data. Ideally, the random placement of traps would create optimal conditions for trap efficacy; however, such ideal circumstances are rarely achieved in practice. Various assessment methods are applied to understand the effectiveness of trap placement. The utilization of traps in planning for pest control has become a widely accepted practice in modern agriculture. It is imperative that this practical application is grounded in real risk assessments, rather than relying on mere suspicions of pest presence or sporadic monitoring efforts that may not provide a comprehensive picture of the pest situation. (López . et al., 2012)

Pest scouting serves as the most extensively employed monitoring tool in agricultural practices, thanks to its broad applicability across various locations and the minimal costs associated with its construction. The detailed scouting procedure encompasses a range of key activities, including diverse sampling methods, effective scouting techniques, accurate determination and thorough documentation, as well as diligent pest reporting, all while often taking into consideration the potential presence and impact of natural enemies in the ecosystem. For growers, the simplest and most straightforward approach to pest scouting involves performing direct visual inspections of their crops. However, the manual identification of pest species and their respective densities can become quite cumbersome and labor-intensive, particularly due to the extensive size of agricultural fields. Additionally, the phenological development of crops often plays a significant role in correlating with the risk of pest presence, making timely assessments crucial. Furthermore, the challenges posed by climate change introduce additional complexities into the realm of pest monitoring, as these environmental shifts can lead to rapid changes in pest dynamics that often exceed the traditional phenological observation scale. Thus, the importance of adapting scouting strategies to these evolving conditions cannot be overstated.

3.4. Pest Forecasting & Risk Assessment Models

The development and the subsequent deployment of sophisticated models that deliver pest forecasts, whether they characterize technical prescriptions or produce indicative risk maps, is arguably among the most significant and formidable challenges faced by researchers and programmers working in the field of pest management. The predictions made by forecasters serve as the ultimate outputs derived from a combination of monitoring data, extensive modeling efforts, in-depth computer analysis, and the meticulous construction of a comprehensive knowledge base. Forecasts that accurately provide pest risk assessments would not only facilitate substantial savings, often amounting to large sums of money, through reductions in both pest populations and the usage of pesticides but would also contribute to a more secure, healthier, and sustainable environment that would be eagerly welcomed by various stakeholders including farmers, environmentalists, and consumers alike. Nevertheless, it is worth noting that the predictions generated by these current models often tend to be less reliable and accurate in a multitude of scenarios, which ultimately results in them being less widely accepted and consequently utilized in practice.

For a significant number of insect pest complexes, researchers currently possess a limited understanding of how global climate change is influencing pest population dynamics, their phenological patterns, and overall risk assessments. The predictions regarding insect pest risks during various times of the year or concerning the potential accumulation of pests in newly introduced areas or with the introduction of new plant species are notably rare. This lack of information and predictive capability is a critical issue that extends beyond just insect pests; similar challenges would also affect models concerning plant diseases and invasive weeds. Indeed, addressing these pressing research questions proves to be considerably more complex and demanding than merely developing the mathematical models that are frequently reported in scientific literature. Forecast models and techniques of this nature could ultimately serve a greater purpose and become more appealing to wider pest management communities if further refined and validated. Nonetheless, it is essential to recognize that some specific focusing techniques may not be extraordinarily innovative but could require incorporation into existing models for better effectiveness. Recent advancements in the realm of forecast models of this type are succinctly reviewed and discussed in a manner that underscores their significance and potential implications for pest management strategies moving forward.

Forecasts characterize technical prescriptions for action. They seek to prescribe resource management actions designed to bring a nuisance species's population down to non-nuisance levels, through resource evaluation and intervention treatment selection. Creating prescriptions generally relies on earlier forecasts modeling population dynamics but is built on earlier action-based decision theory. At earlier stages, focus efforts were made on estimating utility functions, and knowledge about resource action responses was of a more interdisciplinary nature. A more generic discussion can be found on individual versus collective control actions (E Z Tonnang et al., 2017). As treatments relent, follow-up assessments of whether expectations were met are also important products of this modelling approach. Because this elicitation of belief starts with a probability assessment, it is more similar to the risk analysis assessment process used in risk forecast actions. As a result, dealing with a nuisance species's possible action consequences often evolves around belief prompting information, giving rise to models that characterize subjective assessments (Takeuchi et al., 2023).

3.5. Common Pest Identification Guide

Everyone is subject to insect damage. Some years they're a nuisance, other years they're a disaster. The degree of damage that should be considered acceptable and the pest population level warranting action vary with each person and situation. Often cultural practices, resistant varieties, positive and negative natural control, weather, and time factors are unpredictable. To assist in decisions regarding management of insect pests, guidelines, and suggested courses of action have been rated according to their probability of success. This will not eliminate insect losses but may minimize them, provide a more equitable economic return, and explain some of the resistance factors. Follow instructions on treatment guidelines and host plant pest tolerances closely. Modify selections with local experience and weather conditions. Attach special urgency to treatment recommendations on host plant insect tolerances and infesting pests. (J. Wright et al., 1991)

In this comprehensive guide, certain specific insects, along with notable groups of insects, are highlighted as pests that pose significant threats to corn cultivation. These troublesome pests are methodically categorized into two distinct groups: (1) those that are known to affect established corn, whether it is in the seedling stage or as perennial plants, and (2) those that specifically target corn during its vulnerable seedling state. On the left side of the page, readers will find representations of the adult forms of these insects, while on the right side, the larvae or nymph stages are depicted. All illustrations are rendered to scale; thus, the size of each insect is relative to the others presented in the guide. Additionally, general notes regarding the behavioral habits of these insects, as well as management strategies for dealing with these pests, can be found on the accompanying pages of the guide. In particular, special attention is given to the management of the European corn borer, with a separate section outlining potential control measures that can be taken. For the latest and most accurate information concerning these pests and the insecticides available for their effective control, it is advised to contact county Agricultural Extension Service offices, where expert advice and updated resources are readily available.

Part II: Core IPM Strategies

Chapter 4
Preventive & Cultural Control Methods

Contents:

4.0. Introduction

The management principles are common to all conditions. The first approach is to include preventive and cultural control methods into the pest management system. The second method is to select pest-resistant varieties and cultural practices that make it difficult for a pest to thrive. It also needs to prepare options for biological and other appropriate control methods to use if pest outbreaks occur. The final principle is to integrate the various measures selected so that they complement and support each other in controlling the pest with minimal negative impact (Saha et al., 2016).

Cultural controls, including crop rotation, planting site selection, and soil preparation, are effective IPM measures because they also affect other crop insects and diseases. By controlling weeds prior to planting, pest populations already established in the weeds or in the soil can be reduced. Modifying irrigation schedules can promote a dry crop environment making it more difficult for diseases to develop. Soil preparation techniques such as moldboard plowing and harrowing can remove crop residue and bury the soil, both of which could destroy insect, weed, and disease pest eggs, larvae, and spores (Suzuki, 2012).

One of the most effective control measures for soil-borne insect and disease pests is to select planting sites that are free of the pest. A clean planting site assures an insect pest-free area in which to plant, thereby reducing the risk of pest infestations. Alternative planting site options include destroying potential habitat, and using physical barriers such as insect netting and/or fencing.

4.1. Crop Rotation & Intercropping for Pest Disruption

Crop rotation in agriculture is the systematic alteration of crops grown from one season to the next. It is one of a group of practices termed cultural controls that have been recommended as environmentally sound methods for managing insect pests and disease-causing organisms. Crop rotation has traditionally been considered the best tool for managing many soil-borne pathogens of crops. Although cultural controls are widely recommended, data on their pest management effectiveness are generally scarce. This may not be a serious concern with respect to practices such as destruction of debris and other cultural practices that are well known to disrupt pest life cycles. However, rotation has proven more difficult to research and appears to be less understood (J Wright, 1995). Insect pest control by rotation may work by limiting the food supply in the growth cycle of the pest or by reducing survival in the off-season. Research indicates that it is easier to manage a pest that is largely present above-ground than one that has a well-concealed component of its population below ground. For example, downward movement in the soil profile prior to planting corn allows cob borer moths to lay eggs in corn without the intervening winter months exposing them to freezing temperatures. On the other hand, pest insect control with rotation is more easily demonstrated in cases where crops have widely different planting/harvesting periods than with crops that are planted/harvested at similar times. For example, soybean rotation prevents white grubs from getting a foothold, but rotation to barley or small grains doesn't really limit the big cutworm complex which is primarily an above-ground pest. More than 750 insect species in North America are considered agricultural pests and of these about 25 are considered particularly serious pests. Some important insect pest species were omitted from consideration, particularly die-hard pest problems in orchard or vegetable crops. It is hoped that this admittedly crude concept of which pests are likely to be most problematic will be expanded with the input of agronomy department researchers and agronomists with practical experience in the state.

4.2. Sanitation & Field Hygiene Practices

Overwhelming evidence indicates that poor sanitation and hygiene practices are critical first steps in the establishment of an infestation, in terms of both the movement of prone insects to a new field and the initial establishment of that infestation (R. Morrison et al., 2019). Good sanitation deprives insect pests of food, shelter, and moisture, making microhabitats in which pest infestations can hide, feed, and reproduce almost impossible to find. While no one sanitation task can bring a mismanaged pest population under full control, however, a combination of several sanitation protocols used together can be very effective. Intensive sanitation and hygiene practices impose stressors which can impede movement, ability to feed and reproduce, and strength of infestations. A comprehensive assessment of sanitation strategies likely to be effective against the target pest species, the full range of food products, potential habitats for infestation, and time invested in cleaning and similar conditions is essential.

The role of a sanitation protocol as a core strategy is strongly influenced by the sanitation improvement that can be achieved and the existing level of pest populations and infestations. As initial pest populations and infestations start to decline, the threshold at which sanitation procedures would be considered part of integrated pest management (IPM) would also decline. Control would be considered at the point at which expectations for pest control, express as numbers of insects or estimated level of infestation, can no longer sustainably be potentially met by sanitation. If sanitation has improved already reasonable pest control expectations, and no new pests have entered, an IPM approach combining sanitation and monitoring methods may be taken. This makes good sanitation and hygiene practices even more important. They increase the efficacy of any pest management operation, ensure that infestations are prevented from becoming larger than necessary, and prevent exposure to an environment in which infestation is probable.

4.3. Trap Cropping & Barrier Systems

Trap cropping is a method of pest control in which a crop that is more attractive to pest insects as compared to a general cropping system is grown in proximity to the general crop in order to keep the insects out of the desirable crop. In some implementations, the trap crop can be planted first in order to attract pests early. A plant's attractiveness can depend on a number of factors, and trap crops can be chosen to provide desired attractivenesses. For example, a trap crop can be a hybrid plant that contains traits from the general crop but is still more attractive to the pest insect. Trap cropping has recently been proposed for use with pollen beetles in winter oilseed rape . Pollen beetles are able to destroy a crop by consuming pollen and rendering it useless for oil extraction. A method using a trap crop of late-fowering OSR or other crops has been developed that is able to eliminate pests prior to flowering and keep crops pest-free.

Barrier systems are fences or systems of vegetation planted around the crop in order to hinder pest movement into it. These barriers can be made up of pest deterrent plants, plants that provide shelter to pests, or vegetation made up of a number of layers, some short but dense, others tall but sparse. Unlike trap crops, which must be planted extensively, barrier systems are able to be used more easily in a row-crop establishment. Barrier systems can be built simply by leaving a single row of vegetation around the desired area, increasing in complexity as desired. Careful design and placement of a barrier can maximize protection from pests. But on the other hand, barrier systems can also provide useful habitat for pests, potentially worsening a pest problem instead of alleviating it.

Trap cropping and barrier systems are both ecologically sound methods of pest prevention based on farmer experience. However, neither has been carefully studied for their effectiveness or methodology. There are no crop attention prediction models available for either strategy, and there are few if any computer programs that can narrow-down barriers or trap crops to be chosen from or identify configurations of bean rows once a barrier or trap crop position is selected.

4.4. 10 Cultural Practices to Reduce Pest Pressure

Insect pest management using cultural practices requires the same participation by farmers as other pest management strategies. Finding suitable rotation crops may require assistance from several individuals knowledgeable about all aspects of cropping systems. Some rotation crops are known to reduce certain insect pests. Less effort has been devoted to researching rotation crops for control of other groups of insect pests. Rotations that control some pests may not control highly mobile pest species that require more than a single growing season for their life cycle. Decisions regarding program factors such as crop sequence often must consider a number of factors unrelated to pest management.

Research on pest cultural practices is heavily biased toward soil insect pests. The soil stage of many pest species is the least well researched. Acceptable control methods may not be available. The host range and behavior of a soil insect pest may be inadequately understood (J Wright, 1995). Other members of the pest complex may also be difficult to control. Control measures may need to address a species complex, rather than a single insect pest species. For example, a moth pest species may escape control by laying eggs on a plant that is not usually attacked, but a member of the same family still causes damage. This situation is common with the cutworm complex. Pests can adapt to rotations, viewing several crops as "corn," with the northern corn rootworm being a well known example. Crops that do not require insecticides will not be effective rotation crops if pests become tolerant of the insecticides.

Cultural practices in pest management are primarily a preventive strategy aimed at delaying pest establishment and development to ratios that can be economically tolerated. The need for preventive control measures is most acute for pests that cannot be eliminated once present. Implementation of cultural practices may require extensive planning and changes in management before pest levels are known or greater concern is generated, thus demanding educational programs that emphasize long-term planning. The benefit of cultural practices will be greatest if they are used early and continuously in a pest's establishment phase. If better use of crop rotation in corn production was made in Nebraska, it could reduce crop insecticide use in Nebraska and possibly other Midwestern states. Even with rotation, several pest species could remain. Therefore, no one rotational system will control all potential pests here. Furthermore, new rotational systems might result in greater problems with previously common but currently uncommon pests in Nebraska.

Chapter 5
Biological Control in IPM

Contents:

5.0. Introduction

In the following sections, core Integrated Pest Management (IPM) strategies that are widely applied by numerous farmers and dedicated scientists across different regions and countries are thoroughly discussed and analyzed. A brief yet informative introduction is initially provided for each core strategy, paying special attention to the vital and crucial aspect of pest management through the innovative use of biological control methodologies that employs microorganisms and beneficial agents. This innovative approach has gained significant and growing attention in various nations around the globe, highlighting its importance and effectiveness in promoting sustainable agriculture practices. The strategies discussed will include various techniques and methods that are designed to minimize the negative impact of pests while ensuring that agricultural productivity remains high and healthy.

IPM is a systematic approach to pest management where all possible tactics, including biological control, are implemented to prevent, monitor and control diseases and pests. Chemical pesticides are only used as a last resort. Since January 2014, all EU professional growers are obligated to apply IPM tactics according to EU Directive Sustainable Use, indicating that the use of biological control will only increase. However, according to the previously described definition, only few strategies fall under the term biological control. Coinciding with this sharp increase in the potential for biological control agents, research on efficacies of agents and systems has broadened, addressing non-target impacts, safety to humans, as well as on-farm targets. Most biological control agents

(BCAs) sold in Europe are mass-reared arthropods (predators and parasitoids) released in greenhouses and other crop growth facilities. The latter is known as augmentative biological control. Hymenopteran parasitoids have been used most extensively because they are more specific, preventing unwanted non-target effects. Yet, despite the benefits of IPM and the increased number of BCAs available on the market, biological control of insect pests is still applied on a very limited scale.

Conservation biological control has variable success and it remains unclear which structural measures need to be taken to improve pest management. The use of classical biological control is increasingly considered controversial due to non-target impacts, hindering further implementation. A high percentage of BCAs released to fight against insect pests ends up established. Introducing non-native species can cause damage to native plants and biodiversity loss (Francis et al., 2020). Thus, despite the higher chance for success as compared to conservation biological control, this strategy has been banned in many countries. Nevertheless, some prominent classical biological control agents are sold and used. There is an increasing demand for effective BCAs, especially against hard to control pests. However, only a limited number of microorganisms like entomopathogenic fungi, bacteria, nematodes, and viruses have been used for the management of insect pests (J Wright, 1995).

5.1. Biological Control – An Introduction

Pest species occur in disproportionate numbers and are often devastating to plant health and growth. Potential pest species are considered invasive until they are no longer capable of being contained or eradicated. They increase in number exponentially when they find a habitat that meets their biological and ecological needs. Pests have natural enemies, usually specialized predators and parasites that prey on them, but if they reach the production tillage phase of life, the infestation often occurs too rapidly for natural enemies to keep pace with the pest increase. Biological control is the conservation of these naturally-occurring predators, parasitoids, and pathogens or exotic natural enemies that may be released for pest control.

Conservation Biological Control is a way to set the stage for natural biocontrol to occur. It should include a detailed plan that at least considers a predator or natural enemy for biological control prior to any pest infestations. Exotics would be obtained at this stage only if nothing can be found that is already present in the area that can assist with control. This would include searching countries of origin for natural enemies if they can be obtained legally. The strategies considered as options to influence biological control will be species or tactic specific. General categories include providing necessary habitat and host resources for natural enemy populations to survive and proliferate and ensuring that human activities, especially pesticide applications, do not impact natural enemies (J Wright, 1995).

Biological control agents employed in pest management encompass various predatory insects, predatory mites, and parasitoid organisms that work to effectively reduce the numbers of pest populations. Within the field of biological control, there are two primary strategies that practitioners often utilize: conservation biological control and augmentation biological control. Among these approaches, conservation biological control is frequently considered the most advantageous initial option because it tends to be the simplest method to implement and typically incurs the least expense. Nonetheless, it is essential to have a solid understanding of the specific behaviors and life cycles of biocontrol agents, including how they reproduce, disperse, search for their prey, and develop through different stages. This knowledge is critical to successfully design and implement effective conservation biological control strategies that can enhance pest management efforts. (Francis et al., 2020)

5.2. Conservation of Natural Enemies (Predators, Parasitoids)

Conservation biological control aims to promote activity and efficacy of resident natural enemies to enhance biological control services. The prime attractions of this approach to pest management are its reliance on the endemic natural enemy fauna, reduced costs and effort in biological control agent introduction and release, and ecological sustainability. Conservation biocontrol through cultivation of pollen/nectar-providing flowers can enhance natural pest suppression across a wide array of agricultural systems. The system is most effective when widely adopted across regions of intensive production, thereby positively influencing natural enemy communities in all adjacent crops (A. Cloyd, 2020). The system requires further attention to the benefits to pest suppression and economic returns to producers in conventional systems, and to investigate the relative contribution of components of flower strips to pest suppression. Some native plant species with promising traits warrant further investigation of their applicability to the pacific northwest.

In agroecosystems, natural enemies are usually less abundant and effective than their pest prey, complicating conservation biocontrol implementation. As a first step in conservation biocontrol planning, an agroecosystem scientist needs an understanding of the taxonomy, ecology and natural history of natural enemies. This includes a knowledge of basic taxonomy and nomenclature, description of life stages, types of life histories, examples of trophic levels, preferences for prey and habitat types, behaviour, movement patterns, emigration and immigration dynamics, and natural enemy pest prey population models as a function of natural enemy density. A broad knowledge of candidate natural enemies is required, their identification and taxonomy, preferred prey and plant hosts, seasonal activity patterns and food webs. The biocontrol scientist should understand policy issues, crop management practices and systems common in regional agriculture. An essential consideration in this context is the level and types of agricultural practices, including unresolved pest outbreaks, growers' land use intensity, input use levels and IPM adoption. Candidate natural enemies should also be assessed in reference to these considerations. For example, knowledge of land-use intensity, cropping practices and policy issues is vital in assessing the potential conservation of barn owl and the rare cactus moth as biocontrol agents of rodent and cactus pests, respectively.

5.3. Augmentative Biological Control (Mass-Releasing Beneficials)

Augmentation is considered by many people as biological control, which is regarded as buying and releasing beneficial natural enemies. This technique relies on a variety of commercial products, which may be frequently advertised in the mass media. Augmentation is the approach most people think of when they consider biological control. However, it is the least sustainable of the three general approaches to insect biological control because it requires the regular purchase of biological control products. Nonetheless, it can be a particularly efficacious, cost-effective, and environmentally sound approach to pest management when it is applied to situations where it is appropriate. The practice involves the periodic release of natural enemies. This release is based on the idea that in some situations there are not enough natural enemies for optimal biological control, but these numbers can be increased through releases. These practices require a readily available source of large numbers of natural enemies, which has led to the development of a new industry producing and selling these organisms ((J Wright, 1995)).

Two general approaches to augmentation have been widely adopted in practice: inundative releases and inoculative releases. Inundation involves the release of large numbers of natural enemies for the immediate reduction of pest populations. Typically, biocontrol agents are released in numbers many times greater than the target pest population. The expected outcome is immediate pest control. This corrective measure is appropriate in specific situations, such as in greenhouse applications. Inoculation involves the release of smaller numbers of natural enemies at various intervals to attempt to maintain low pest populations. Releases typically begin at pest population densities that are very low and below the economic injury level. The expected outcome is to keep pest numbers low. This preventive measure is not recommended unless the preceding conditions are satisfied.

5.4. Microbial Biocontrol (Bt, Fungi, Nematodes)

Microbial pesticides are the microorganisms or their byproducts that are helpful for pest and disease control in the environment. Microbial products have contributed more than chemical pesticides for mankind since they are eco-friendly. The microbial control of insect pests has gained considerable importance in the past few decades due to the harmfulness of synthetic insecticides to humans, animals, crops, and environment (S Vimala Devi et al., 2012). Microbial pesticides, which include naturally occurring pathogens like bacteria, fungi, and viruses, offer an alternative to chemical pesticides. The success of microbial pesticides is based upon: safety to humans, animals, and environment, persistence in nature, cost-effectiveness, target specificity against pests, and less risk of developing resistance in pests. Microbial insecticides are crop compatible and can be used in IPM program with minimal risk of development of resistance compared to chemical insecticides as the rate of application is usually very low and target specific in behavior. Fungal pathogens such as Beauveria, Metarhizium, and Isaria and bacterial pathogens such as Bacillus thuringiensis were developed into commercially available formulations for insect pest control. Except for a few, the efficacy of other microbial pesticides are low compared to chemical pesticides. Most of them are still in the experimental phase of their development and not commercially available. There is a need to develop strategies for distribution, implementation, and use of existing microbial pesticides. By taking into consideration the vigorous pest incidence, chemical pesticides will be banned for commercial use. With the development of microbial pesticides and improvement of bioformulations the pest management will be possible with fungi, bacteria, and nematodes at the forefront of pest management strategy. The exploration of infectivity, virulence, persistence, inducing resistance, bio-formulation, and mass production of bacterial, fungal, and nematode pathogens against insect pests will be the future research areas. During the coming decades, biopesticides based on microbial pathogens will play a major role in insect pest control.

5.5. Ladybugs in California Citrus – A Classic Success Story

Citrus in California cultivates about 100,000 acres with an annual gross income of $600 million. The greatest pest threat is the aphid, Toxoptera citricida, that feeds on new growth, causing curling and yellowing of leaves as well as loss of productivity. In addition, Toxoptera pushes a large quantity of honeydew which causes sooty mold that prevents sunlight from reaching foliage. Initially, Toxoptera was controlled by a single plane release of the parasite, Lysiphlebus testaceipes, that was able to reduce the aphid population by 95 percent. However, L. testaceipes was unable to control Toxoptera in mature groves and the need for a pesticide was seen. Because of the heavy use of broad-spectrum insecticides, the number of ladybugs was reduced and Toxoptera and its parasites were able to build up to damaging levels. This case documented how ladybugs were brought back to citrus by habitat manipulation to re-establish the biological control. On 17 March 1972, a ladybug habitat was created by releasing seedlings from an overlooked site that had been uncut for 10 years. Yield from these trees was expected to be five to six tons of fruit per acre and to be increased. Since many adult ladybugs were present at the start of the study, restoration of the biological control was rapid. As the ladybugs built up in numbers and dispersed to the groves, the Toxoptera numbers fell to low levels. There were no longer any aphid infestations and a better fruit quality was obtained. A one-year jubilee celebrated the return of the ladybugs and the reestablishment of the biological control. Pesticide use was reduced by 90 percent resulting in an annual savings of $250,000 for the entire operation. Also, by not using pesticides, the natural cycle was allowed to establish and plants were not damaged (Demirozer et al., 2012).

5.6. Red Palm Weevil in Date Palm

The red palm weevil, scientifically named Rhynchophorus ferrugineus, is recognized as a major pest problem that significantly affects the cultivation of the date palm, known as Phoenix dactylifera. This tree is cherished in the Middle East and North African countries, celebrated for its incredible nutritional value and its cultural significance as one of the most-loved fruits. The infestation of the red palm weevil has led to substantial cost overruns in controlling this pest, particularly within the ornamental and date palm cultivation industries across various nations. Reports indicate that the date palm faces threats from numerous insect species, which include not only sap suckers but also leaf miners, black scorch beetles, and trunk borers. Despite this array of pests, the red palm weevil is considered the most critical threat, highlighting the urgent need for active endeavors aimed at controlling its population before the problem escalates further. (Nurashikin-Khairuddin et al., 2022)

The economic importance of date palm is significant and multifaceted, reflecting its immense local, national, and international value across various sectors. The exportation of date palm products earns vital foreign currency for producing countries, highlighting its role in global trade. Annually, approximately 5 million tons of dates, which are valued at a remarkable USD 2.5 billion, are exported to international markets. Beyond mere figures, date palm thrives as a source that produces an impressive array of a thousand different products, thereby contributing not just to the economy but also to essential nutrition in the countries where it is cultivated and harvested. The popularity of dates and their various derivatives resonates with people of all ages, making them a staple in many diets around the globe. Additionally, the wood derived from the date palm is utilized for construction purposes, offering substantial credit to its mechanical strength and durability. The rough fibers from palm fronds are ingeniously employed in the making of brooms and carpets, showcasing the versatility of this plant. Moreover, palm leaves are cleverly repurposed for roofing and various other construction applications, serving both functional and aesthetic roles in residential and commercial buildings. The midribs of the leaves are skillfully transformed into a range of household articles, further enhancing their utility. The thick, fibrous bases of leaf stalks and spathes are also ingeniously crafted into products like carpets, ropes, and baskets, demonstrating the creative uses of every part of the date palm. Moreover, resourceful farmers collect water vapors from the petiole and spathes to extract sugar, tapping into another aspect of the date palm's economic potential. The plant is also known to produce essential soil nutrients, edible products, and traditional medicines, making it an invaluable asset in the agricultural landscape. Overall, the contributions of date palm extend far beyond its fruit, playing a critical role in the livelihoods of many and reflecting its diverse economic significance.

The majority of date palm plantations around the globe are currently under significant threat from the Red Palm Weevil (RPW), which has reportedly infested and adversely affected around 40 species of palm trees. The damage caused by this notorious RPW infestation has attracted considerable worldwide attention, especially as its rapid spread has become increasingly alarming in various regions including Asia, Europe, and the Americas. This highly invasive pest has led to the culling of thousands of disease-infested palms across many countries, with Iraq, Saudi Arabia, Egypt, Oman, the UAE, and several others being particularly hard hit by its detrimental impacts. In addition, the additional costs incurred due to the RPW infestation, particularly in terms of implementing other pest control measures, are estimated to reach around 4 billion dollars on a global scale. This figure is not only staggering but is also rising rapidly as new infestation hotspots are continuously being detected in an ever-expanding number of countries. Consequently, intensive management strategies have been adopted in numerous front-line states to effectively combat the spread of RPW and maintain the pests under strict control, demonstrating the urgent need for global cooperation and action in addressing this relentless challenge.

Chapter 6
Mechanical & Physical Controls in IPM

Contents:

6.0. Introduction

One method of managing pests is to exclude them from the crop and prevent them from establishing. This can be achieved by ensuring that greenhouses are adequately sealed with only the required doors and vents, and that they have screened windows. Safer horticultural buildings can also be made to exclude pests by ensuring that any cracks or holes in the structure are sealed, and that there are no holes in the plastic or mesh coverings (Suzuki, 2012). Procedures to remove pests from the production area are also important, but may be difficult to present. Under some constraints, a sweep net, moth traps, or traps baited with sticky glue or pheromones can be used. To eliminate aphids and other soft-bodied insects, an insect vacuum can be developed that uses a low-pressure fan to draw insects into a mesh net.

Mechanical methods are also used to destroy pests. For example, aphids can be killed instantly by blowing air developed by a blower or a hair drier. However, the use of strong air currents could also destroy the leaves. Smoke, hot air, or water vapor have also been used to heat greenhouses and destroy pests. However, this heat may have a negative effect on crops.

Physical methods control insect numbers without the use of toxic substances. This achieves a metabolic disturbance in the insect, inhibiting development and reproduction, and eventually killing it. Temperature, radiation, moisture, oxygen, or ultraviolet art are all forms of energy that have been, or can potentially be, used for this purpose. The damage inflicted on the pest by these energies may not always be lethal. There is concern with temperature control as a disinfestations procedure, as high temperatures must be maintained for several hours, and pests may become more active and move to escape. The efficacy of this procedure may also be influenced by the physiology of the pest, the desired crop, and other inherent variables (Francis et al., 2020).

6.1. Mulching, Row Covers, and Exclusion Nets

Within the context of monotonous horticultural production systems, conservation biocontrol practices are more likely to occur at the micro-level (biodiversity at field and farm levels) than at intermediate (landscape and community levels) or macro-levels (regional and national legislation) because they are less costly and may provide farmers with higher pest control services without the need of expensive institutional support (Kuesel et al., 2019). Consumers play a significant role in the main aspects of horticultural production systems involved in biocontrol and in the management of pests. An increasing amount of environmentally-conscious consumers favour honey or linden-flavoured apple trees and the non-cultivated plant cover of early-spring flowering herbs, as they are important resources for bees. The importance that avocado consumers afford to their health also influences the degree in which Brassica weeds are controlled (Candian et al., 2021). Some biocontrol practices occur at the micro-level and are based on exclusion nets to control sap-sucking pest species, which are a constant problem for nectarines. The results of a horticultural system, which used exclusion nets to protect nectarine trees from the pest Myzus persicae and other harmful insects, were evaluated in comparison with trees uniformly treated with the acaricide. In this context, exclusion nets reduced pest populations and increased fruit production, which were beneficial for marketability and quality. The apples produced were sold in high-quality levels and under €0.40 kg−1, whereas lower prices were paid for the non-netted trees' fruits. The use of exclusion nets leads to a general reduction in insecticide treatments and to lower concentrations of pesticides in fruit. In addition, no negative effects on postharvest rots, fruit quality, or nutraceutical properties were recorded.

6.2. Thermal Methods (Solarization, Flame Weeding)

The utilization of heat to manage pests is one of the oldest approaches in agriculture (Suzuki, 2012). For example, thermal methods help clean soil in glasshouses of weeds and soil-borne diseases. Popular thermal methods include soil solarization, hot-water treatment of seeds, and flame weeding. Natural enemies that are susceptible to heat have been released during cool or rainy weather. In wide tropical areas, soil solarization has strongly reduced the numbers of various soil pests in various fruit, flower, and vegetable crops. Both manual and automated flame-weeding have been introduced in greenhouses and vegetable fields to reduce the numbers of various annual and perennial weeds. Several thermal methods currently under investigation are expected to support sustainable agriculture.

Soil solarization is the most effective and environmentally friendly thermal method to disinfect soil. In humid coastal areas, it has been used to reduce the numbers of various soil-borne pathogens, nematodes, and various weeds in various fruit, flower, and vegetable crops planted under plastic greenhouses or tunnels. Because biogas digesting of various agricultural and animal waste produces noxious materials, climate control is required and fertilizer is applied. While chemical pest management methods have inevitable harmful side effects on the soil environment, as soil sanitizing technologies, UV radiation from artificial light for pest initiation control and soil solarization for soil disinfection are expected to facilitate fertile and sustainable agriculture. Regardless of any pesticide, prevention against pest and disease occurrence should be focused on.

6.3. Vacuum & Trapping Systems for Large-Scale Farming

Vacuuming systems have been developed for the field control of a number of insects from diverse orders including Lepidoptera, Coleoptera, Hymenoptera, Diptera, Hemiptera, Orthoptera, and Arachnida; it has been demonstrated that, depending on crop and insect species, up to 97% of the pests can be removed from a 100ha field in a single day. Furthermore, trapping of male moths can provide near-perfect control of lepidopteran pests, while the vacuuming of the crop canopy can eliminate the majority of female moths. Up to 90% of the eggs of certain pests can be removed from leaves by the mechanical approach of a 'hot wax' method, and the hatchability of the remaining eggs is greatly reduced. Currently installed systems include orchard vacuums to remove fruit flies, viticulture vacuums to collect and destroy grape berry moths, and large-horsepower tractor-drawn vacuuming systems for truck cropping insect control. Commercially available systems include vacuum systems to suck crop pests from the air and those attached to a tractor for sucking them from the crop. Vacuum systems to suck insects off trees and vines have been tested in vineyards, fruit orchards, and nut orchards in various countries. Nets are used to close the sides or ends to exclude insects from entering a crop but allowing beneficial insects to exit if they want. This technique is most effective early before pest populations build up; however, it does require a net to be purchased and installed for which installation might require the use of cranes. Fermenting compost has been found useful in attracting certain fruit fly pests to areas outside vulnerable primary vegetable crop, and this approach can potentially be used for a wide range of other insects. The trapping can be differentiated between two types: The first type was the tracking of male moths and females of a few other species with the potential to develop the systems. Tracking with a bait and ducted fan system has been developed and used commercially; funnel traps and unbated toothbrush design swatters have been used satisfactorily in a few crops. Enormous strides with similar designs have also been made by a few enterprises. The second was the tracking of all insects. The dynamics have been examined and currently, designs include the inside and outside entrances and an ultra-quiet vacuum and elevator type catcher to avoid adults traumatized with air bursts hitting the sides of the doors. This system can be used in open fields and, if fully equipped and replicated, measure all insects in a region.

6.4. DIY Pest Traps for Smallholder Farmers

In the past few years, there have been numerous publications detailing the construction of inexpensive but effective insect traps that have been specifically devised and scaled into the millions based on the utilization of readily available local materials. Such plume-based mass trapping devices essentially rely on ambient wind acting on their plume diffusers to effectively capture insects, which consequently precludes either hurricane or near calm conditions, severely limiting their overall effectiveness in those specific scenarios. In windy conditions, plume-based trapping efficiencies often decrease significantly, while the traps tend to remain effective at night when insects are more active, which may inadvertently lead to pest population explosions during the day when these traps are less operational. This inconsistency presents challenges to pest management strategies that depend on these traps for controlling insect populations.

To address this resource gap, a new model for a foldable fan-trap is described here; an inexpensive, compact, and open-source device option for insect monitoring that requires no readily available local materials, tools, or skills to fabricate. It plausibly projects much wider usage by growers in the developing world, one of the UN Food Systems Summit 2021 Gamechangers providing cheap traps from plastic waste streams monitored with inexpensive cameras and open-source sensors and agri-advisory software central to IPM extension services. Initial pilot studies conducted in Jade City, a fabled gem-rich economic transfrontier zone in south-west China, and other regions brewing with traps using standard aluminium or plastic foil, indicate remarkably fast and easy assembly and impressive trap capture efficiencies. An inexpensive 'DIY' Rosa-Fan trap is proposed as a complement to (Grégoire et al., 2022) and numerous self-publicised gallery traps, megabauds, and rod-style motors for trapping insects of many sizes in any direction without bias or emissivity constraints.

Chapter 7
Chemical Control in IPM (The Last Resort)

Contents:

7.0. Introduction

In IPM, chemical control should be the last resort, after all other methods have been tried. Chemical control also includes control by harming the environment. Pesticides cause major outbreaks of spider mites by eliminating natural enemies, and furthermore, they are more harmful to predatory mites than herbivorcs such as Tetranychus urticae. Accumulation of knowledge on side effects and methodologies to test for this is required. Development of chemicals that selectively target herbivores is also necessary for the shift from traditional pesticide use to IPM (Suzuki, 2012).

Most allelochemicals produced by plants, microorganisms, and carnivores are not used for commercial purposes. These include the antioxidative luteolin, jasmonic acid, and galloylglucoside, which induce plant resistance against herbivores. It is important to accumulate knowledge on mechanisms and pathways of action to encourage the commercial production of these. Biological control has been widely adopted (A. Umina et al., 2015). Progress in the rearing and commercialization of natural enemies of mites and insects is commendable. The next step is establishment of a social system that reinforces the adoption of biological control, including followed-up experiments and collaboration between farmers and research institutes. In particular, it is necessary to consider the processes and time needed for natural enemies to proliferate and control pests.

Planning the reinforcement of biological control should not be based solely on a week's time scale with the idea of the withdrawal of pesticide use.

7.1. Selective Pesticides & Reduced-Risk Chemicals

All approaches that increase the ecological selectivity of pesticides should indeed be carefully considered and thoroughly examined. This aligns well with the conclusion drawn from section 1.2 and is grounded in the fact that several of these methods are already widely recognized and implemented in various agricultural practices. Numerous approaches have been described in detail, yet it is important to note that for some of these methods, there exists little concrete supporting evidence available to back their efficacy and benefits.

Integrated pest management strategies are designed specifically to reduce farmers' reliance on a limited number of treatments, which could potentially have detrimental effects on the environment. Furthermore, these treatments may exert a significant selection pressure that could lead to the development of resistance among pests and weeds. To illustrate this point more clearly, weed and pest management serve as concrete examples. The central goal of these strategies is to emphasize that the concerns surrounding pest management extend far beyond the narrow scope of chemical approval processes. In fact, a comparable argument could be made regarding the necessity for ongoing and rigorous analysis of manufacturing processes. Such assessments are vital in identifying and mitigating the risks associated with the emergence of new residues, which can adversely affect both food safety and the broader environment. By adopting a holistic approach, integrated pest management seeks to promote sustainable agricultural practices that are less reliant on harmful chemicals, thus ensuring long-term ecological balance and food safety for future generations.

There are two separate, albeit interrelated, issues concerning the development of pesticides and the processes involved in their approval. First, there is a need for all possible approaches aimed at enhancing the ecological selectivity of pesticides. This aspect is crucial alongside the range of measures that are currently addressed by existing pesticide approval processes. Additionally, this matter is relevant to any reviews that pertain to specific pesticide formulations and methods of application. The focus on ecological selectivity plays an essential role in ensuring that environmental impacts are minimized while maintaining efficacy.

Any approach that reduces 'unnecessary exposure' of pests and other organisms to harmful treatments should be considered. There is a need for a more complete agricultural risk assessment of 'reduced-risk' pesticides and/or alternative practices. It is also suggested that broader independent analyses of countries about which pesticides were approved despite a lack of data would be useful.

7.2. Resistance Management (IRAC Guidelines)

Insect resistance poses a threat to the future use of microbial Bacillus thuringiensis (Bt) formulations and transgenic plants expressing Bt insecticidal proteins. Academic and government scientists, public interest groups, and organic farmers have expressed concern that the potential for widespread planting of these genetically-modified plants will hasten insect resistance to Bt endotoxins. In 1999, the American Crop Protection Association announced a commitment to implement resistance management plans for Bt corn. Under the authority of the Federal Insecticide, Fungicide and Rodenticide Act (FIFRA), the United States Environmental Protection Agency (US EPA) is to ensure that there will be no unreasonable adverse effects on the environment from the use of a pesticide, when all economics factors are considered (M. Matten et al., 2004). In 2000, the US EPA imposed new Insect Resistance Management (IRM) requirements on registered Bt crops to combat insect resistance to Bt endotoxins. Sound IRM will prolong the life of Bt insecticides, benefiting growers, producers, researchers, and consumers. Academic and government scientists agree that insect resistance is likely. However, there is disagreement with regard to factors that enhance or impede susceptibility to IRM agroecosystems. Improving this understanding is critical prior to the large-scale application of transgenic technology IRM.

Strategies to address insect resistance to Bt by the US EPA include instituting IRM plans in collaboration with stakeholders and fundamentally understanding how pest resistance happens and how it can be stopped. To date, US EPA has proposed, but is implementing a 50% refuge plan only for Bt corn expressing insecticidal proteins expressed in a single form. Development of sustainable IRM strategies for Bt field corn has required ongoing engagement and collaboration among stakeholders. The US EPA has held multiple fora for stakeholder discussion, research presentations, and working groups focused on insect resistance management for Bt crops. Currently, the NC205 committee formally addresses resistance and other research issues related to the ecology and management of the European corn borer and other stalk-boring Lepidoptera. NC205 has sponsored annual resistance management meetings since 1998 and is used to present research results and responses from stakeholders on IRM issues and to establish research priorities.

7.3. Biopesticides (Neem, Pyrethrins, Essential Oils)

The use of plant-derived materials for pest control is one of mankind's oldest endeavours. Despite their long history of use as insecticides, traditional compounds such as neem oil are often regarded with caution. Eventual mainstream positions regarding these materials for pest control demand extensive research and development. Smallholder farmers can benefit from plant-derived insecticide sprays, especially from plants whose active ingredients lack field-tested studies. Local testing and evaluation of proposed plants should occur in collaboration with smallholder farmers in order to develop methods that are both effective and practical within local contexts. It is in smallholder farming systems that various pest control compounds currently in use have emerged. Despite potential pitfalls associated with extreme simplicity, developments still need to occur via the collation of information from around the world, especially with regards to success stories in similar contexts (V. R. Campos et al., 2016).

The development of plant-based pest management resources can act as a bridge into integrated pest management (IPM) and the use of other subtle materials and methods, being one of several steps required to engage smallholder farmers in more sophisticated alternative pest management strategies. Traditional plant-based insecticides could be refined and further developed to make them eminently more effective and modify their modes of application further. Pest control systems that are currently sub-optimal could be improved, giving local systems legitimacy and encouraging their continued use. A broader engagement with natural products-based pest management is required with active attention drawn to subtle options and their investigation as pest management methods in farming systems. At the same time, the limits of traditional botanic insecticides as pest management methods should also, where warranted, be integrated into the research agenda (Mandela Fernández-Grandon et al., 2020).

7.4. Factors Affecting Application and Chemical Deposition

When applying pesticides there is always a chance some will escape from the target area. Inefficient application has long been an issue facing the pesticide application industry. Drift is of concern because it removes the chemical from the intended target making it less effective and the chemical is deposited where it is not needed and often not wanted. The second concern is generally the most critical because the pesticide becomes an environmental pollutant in the off-target area. Off-target deposits can injure susceptible vegetation, damage wildlife, and contaminate water supplies. Costly problems can result when carelessly applied pesticides, especially herbicides drift and cause damage to economically or aesthetically important crops. Although drift cannot be completely eliminated, the use of proper equipment and spraying techniques will maintain drift deposits within acceptable limits. The primary recommendation for drift control is to read the pesticide label. (E. Wolf, 1996) There are two ways that herbicides move downwind to cause damage: particle and vapor drift. Vapor drift is associated with the volatilization of herbicide molecules and their movement off-target, making it independent of the application. Particle drift is the off-target movement of spray particles formed during application. The amount of particle drift depends mainly on the number of small "driftable" particles produced by the nozzle. By varying the nozzle design parameters, such as size, orifice type, mounting angle, and swirl chamber length, different droplet size distributions can be produced. In general, smaller droplets are more likely to drift and the right nozzle must therefore be selected depending on the type of pest being treated, target site, application conditions, pesticide formulation, equipment, and the upcoming weather. (Chen et al., 2024)

7.5. Calibration of Aerially Applied Sprays

Calibrating your spray equipment means adjusting your sprayer so that the amount of chemical you want to apply to your crop and the amount that is actually applied to the crop is the same. Your equipment is not calibrated if in the middle of applying an insecticide on your crop your spray tank is empty. Likewise, your equipment is not calibrated if you sprayed your whole field and find half the tank still filled with tank mixture. Once your equipment is calibrated, you can use it to deliver accurate amounts of chemicals on any size area. You will know how many gallons of tank mixture you would need to spray 1 acre of your crop. For example, suppose the instruction on the pesticide's label says to apply a dosage of 2 pounds of the pesticide per acre but it does not say how much water to mix with the 2 pounds. If you calibrated your sprayer and found that you would need 50 gallons of tank mixture to spray 1 acre, you would know that you should mix 2 pounds of pesticide into 50 gallons of water. You could make just 25 gallons of tank mixture by adding 1 pound to 25 gallons of water. Suppose you calibrate your sprayer and find that you need 50 gallons of tank mixture to spray 1 acre. If your job is to spray half an acre, you can estimate that you need only 25 gallons of tank mixture. Calibration refers to the process of ensuring that pesticide application equipment is delivering the desired amount of pesticide, either a liquid or dry formulation, to a target area as uniformly as possible (R. Rider & C. Dickey, 1982). Pesticides are expensive, and pest control is one of the largest single costs of production for creeping bentgrass. Pesticides provide effective pest management, but underdoses allow pests to exert damage, while overdoses can result in crop damage and pollution of the environment. Efficient pesticide application requires appropriate sprayer calibration, which provides accurate field performance. Proper calibration of pesticide application equipment is essential to ensure optimal pest control. Proper calibration can reduce pesticide application costs by maximizing control efficacy for a given dose. Uniformity of application is essential for pesticide control. Improper sprayer calibration is a major cause of inefficient application and improper pest control (Arakaki & Nagamine, 2004).

7.6. Common Pesticide Mistakes to Avoid

Integrated Pest Management (IPM) is a systematic approach to controlling pests, both inside buildings and outdoors. To be effective long-term, an IPM program must focus on nonchemical controls—such as sanitation, mechanical control, and the selection of pest-resistant varieties—while also incorporating the selective use of chemicals only when necessary. Because they are closely related, horticulturists and landscape managers often choose to integrate their pest management programs for both plant and pest problems.

Commercially available pesticides are not required to carry disposal directions on the label since they are not considered wastes until empty or unusable. Unlike household containers, all pesticide containers should have disposal instructions. To avoid contamination of local land and water, disposal instructions should be followed, and any product that is not usable should be returned to the dealer for proper disposal. When recycling is not possible, pursue disposal via a household hazardous waste program. Pesticide containers should be rinsed in a manner similar to household pesticides, disposing of the rinse water on a vegetated area far from streams or ditches, puncturing the container so it cannot be used again, and placing it in the household trash. Rinse water from larger containers should be applied to land where a decision not to plant or to rotate crops is made. However, the local extension agent should be consulted about preferred disposal methods and any special regulations. Always check the label or material safety data sheet for any specific disposal instructions, and do not take anything for granted (Robotham et al., 2000).

Many agricultural chemicals that are not used are especially toxic to people, pets, and wildlife. Repeated exposure, over time or through several routes of exposure, can lead to chronic or cumulative effects. Farm operators using pesticides are more likely to be exposed than others; often their families suffer from the same effects. Transporting or applying pesticides carelessly can lead to illnesses among neighbors. More individuals complain about pesticide symptoms than about symptoms from weeds, garden pests, or structural pests. Hand weeding, hoeing, and mulching can control most weeds, with careful supervision. Bugs can often be removed by hand, especially on vegetables and garden plants before damage is done. Cleaning up dead leaves and debris—where slugs, root maggots, and cutworms are likely to breed—removes potential homes to pests where they cannot breed, eat, and grow. Sticky disks are traps for flying insects around greenhouses or the home, or around individual carrot or cabbage plants. Draw out ants by placing a few tuna fish cans buried in the ground up to the rim and containing a little bait. Household pesticides seldom kill a harmful insect or weed, yet they may injure a pet, cause a rash on a child's skin, or lead to a case of asthma (Grisso et al., 1998).

Chapter 8

Behavioral & Genetic Control in IPM

Contents:

8.0. Introduction

The behavioral control of pest populations effectively exploits the evolved or naturally occurring behaviors that have significant ecological roles. These behaviors include the use of aggregation pheromones as well as anti-aggregation pheromones, which play a vital role in restoring family-level kinship within populations. This approach to pest management applies broadly across various taxa and different life stages of organisms, encompassing behaviors such as oviposition or rcsourcc selection. By influencing these behaviors, there is often a reduction in subsequent damage that might otherwise have been caused by these pests. This field of study also intersects with chemical ecology, exploring how quasi-social behaviors exhibited by plant pests can inform control strategies and lead to the development of non-chemical methods of repellence that serve as effective deterrents against pests. Behavioral control tends to target specific pest species or classes somewhat imprecisely, yet it capitalizes on fundamental natural behavioral responses instead of relying on more aggressive attract-and-kill strategies. This group of behaviorally based methods is especially relevant in situations employing invasive organisms where the importation of natural enemies or other controls for pest organisms is hindered due to various biogcographic or regulatory challenges. As such, leveraging natural behaviors presents a sustainable avenue for managing pest populations while minimizing environmental impact through chemical means.

Behavioral control must be carefully integrated with a variety of other effective control measures within any comprehensive total management strategy. It is essential to recognize that the complete removal of pests from a given system is generally impractical and often unattainable. Over time, pest populations invariably develop resilience, whether through behavioral adaptations or genetic evolution, allowing them to escape conventional control efforts. In this context, behavioral management must primarily focus on robust prevention strategies and non-involvement, while simultaneously maintaining flexible, quick-reacting measures designed to effectively protect high-value resources that are at risk of pest damage. Furthermore, the release measures aimed at protecting pest control doses might be more susceptible to adjustment, as higher doses could be less affected by inequities in cost-benefit assessments. Importantly, these control efforts should be designed to have a high probability of success. Relying solely on conventional population control methods often leads to increased costs and a greater effort required for enforcement since the pest population tends to rebound dramatically, often leading to delayed or self-inflicted outbreaks that result in unforeseen expenses that can reach significant levels. Therefore, it is crucial to understand that behavioral management must always prioritize the protection of the economic centerpiece in any total management strategy, ensuring that resources are allocated effectively and strategically in order to mitigate risks associated with pest populations. (Harvey-Samuel et al., 2017)

8.1. Pheromones & Mating Disruption – E.g., Red Palm Weevil

Over the last few decades, the widespread use of herbicides has led to a significant increase in herbicide resistance among various weed species. This growing issue has, in turn, made the management of these troublesome weeds much more constrained and challenging for farmers and land managers alike. As a consequence of these developments, there is now an urgent and pressing need for more sustainable alternatives that effectively harness cultural weed control measures. These should be integrated with a range of other eco-friendly technologies to facilitate economically viable and effective weed management strategies. One promising approach that falls within this broader strategy is the use of pheromones. Pheromones are a specific type of semiochemicals that are produced by an individual organism and serve to mediate a behavioral response within another individual of the same species. Their unique properties make them particularly valuable in monitoring and managing pest species that are crucial to agricultural systems, especially due to their high specificity. Pheromones can be utilized in a variety of ways, both in curative and preventive management approaches. For instance, they can be employed to monitor pest abundance, track pest distribution across different landscapes, and effectively trap pests, thereby helping to reduce their populations and manage their impact on crops. (Na et al., 2024)

As of now, a total of nine commercial products that contain pheromones have been successfully launched in the UK. One prominent and well-recognized illustration of this is the method known as "mating disruption" or "sexual confusion," which is primarily implemented in fruit and nut trees. This innovative approach has proven to be extraordinarily effective in combating pests like the codling moth, various species of leaf rollers, and several other significant pests as well. Additionally, the white fly that affects the Casuarina plant is specifically targeted in Nigeria through similar methods. Pheromones serve a dual purpose: not only are they utilized for managing pest populations, but they also play a crucial role in monitoring indigenous pollinators in orchards. This application is an eco-friendly strategy for preserving vital pollinator services that are crucial for the health of ecosystems. Nevertheless, it is important to note that while the prevalence of pheromone-based solutions is gradually increasing, the majority of the plant protection products that are registered globally mostly focus on specific pest species. This observation is rather peculiar, especially when one considers that we have known for many years that pheromones can be synthesized in a laboratory setting. This capability theoretically enables broader pest accessibility, allowing for more comprehensive pest management strategies. Presently, however, the relatively high costs associated with the mass production of these pheromones result in their registrations usually targeting high-acreage industries, such as those involved in fruit tree cultivation and nut crop farming. Other types of crops present more significant challenges. This is primarily due to the lower investment returns they offer, alongside the complexity of the pest species themselves, which are often too polyphagous. In many cases, this polyphagous nature necessitates the use of multiple different pheromonal components that must be carefully mixed together to achieve effective pest control.

The red palm weevil, scientifically known as Rhynchophorus ferrugineus Olivier, represents a particularly devastating pest that inflicts harm on a wide variety of palm species, with a notable preference for date palms. These trees are not only a staple in the agricultural landscape but also hold significant cultural importance in numerous arid nations, reflecting traditional practices and local economies. In the United Arab Emirates, particularly in the vicinity of Dubai, there have been alarming reports of severe infestations caused by the red palm weevil, which poses a serious threat to both agricultural sustainability and the flourishing tourism industry across the nation. At present, the predominant method employed to control the red palm weevil infestations involves the application of chemical pesticides. Unfortunately, these pesticides are known to carry substantial risks, posing dangers to human health as well as to the surrounding environment. Given the rising global demand for date palms, which increases every year, there is an urgent need for the development and implementation of sustainable management tools to effectively combat this pest. One promising avenue in agricultural pest management is the use of pheromones, which are natural chemical signals employed by many organisms for communication. By leveraging these pheromones, it may be possible to manage pest populations in a more sustainable manner. To assess the feasibility of using pheromones in the management of red palm weevil populations, a selection of attractive pheromone compounds was meticulously chosen and evaluated for their efficacy in attracting and trapping adult specimens of the red palm weevil. Through this research, we aim to contribute to a more environmentally friendly approach to pest control.

8.2. Sterile Insect Technique (SIT) – E.g., Mediterranean Fruit Fly Control

The SIT, or Sterile Insect Technique, constitutes a highly specific strategy that focuses on targeting specific species of phytophagous pests. In this innovative method, male insects are subjected to irradiation, a process that effectively sterilizes them while ensuring that their ability to compete for mating remains intact. Once these sterile males are released in large and sufficient numbers, they engage in competition with wild males for access to females. This competitive advantage gradually results in a decline in the local wild insect population. Over the span of more than 50 years, SIT has been successfully employed against a variety of pest species and has seen extensive application against the Mediterranean fruit fly, commonly known as the Medfly, across multiple continents. The overall success of SIT programs is heavily reliant on the timely release of adequate numbers of high-quality sterile males within the designated target areas. The long and rich history of classic SIT demonstrates that effective pest control through the release of sterile insects is only achievable if the sterile males are capable of being detected and subsequently exploited by their wild counterparts. Most successful SIT programs are underpinned by a solid understanding of the population ecology of the relevant insect species, coupled with the implementation of appropriate breeding and release techniques tailored to their specific needs. However, a significant number of SIT programs have faced failures or continue to experience challenges in achieving success due to a range of issues, including the release of poor-quality insects, inadequate numbers, ill-timed releases, or overall improbable strategies. This complexity highlights the necessity for careful planning and execution in order to optimize the effectiveness of SIT as a pest management tool. (Dumont & F. Oliva, 2024)

Wild sterile male releases are often carried out by air- or ground bombardment methods. In bombardment feeding, a food bait enriched with either irradiated males or a mixture of radiated and wild males is broadcasted simultaneously within the area of interest. For insect species with a strong competitive advantage and/or high population growth rates, saturation releases can sometimes be accomplished. For other species, oogenesis competition is introduced, where wild fertile females are fed on toxic bait to drive the natural population to zero before sterile males are released. Baited traps are also recommended for many species, but for some, it may not be effective due to the difficulties of setting traps on nectar plants. Several species lack a suitable bait or trapping method, and finding such a method for these needs to be addressed. When the released wild males are too numerous and/or over a long period, compensated feeding may occur, which neutralizes the wild consumed males' effect. Sterile males also require supplementary feeding to compensate for the sterilization treatment's adverse longevity and nutritional effects.

8.3. CRISPR & RNAi – The Future of Genetic Pest Management?

Integrated Pest Management (IPM) represents a comprehensive and systematic ecological approach designed specifically for the effective control of pests. This method places significant emphasis on the critical importance of integrating a diverse array of environmental controls alongside supplementary pest management techniques as they may be required. Historically, the primary focus of the IPM strategy revolved around enhancing previously established pest control methodologies, actively deepening the scientific understanding of complex pest systems, and delivering suitable, effective pest management techniques that work towards minimizing the overall reliance on chemical pesticides, which can have detrimental environmental effects. In recent years, there has been a notable rise in the acceptance and widespread implementation of IPM practices across various sectors. This shift reflects a broader recognition of their significant value in sustainable agriculture and environmental conservation. Numerous IPM programs are now actively sought after or have been established by both state and federal government agencies, as well as by numerous national organizations operating in a multitude of different countries around the world. Furthermore, International Non-Governmental Organizations (INGOs), along with significant charitable foundations, frequently support these vital initiatives, particularly in the context of foreign aid deliberations and development programs aimed at improving agricultural practices in developing countries. These IPM programs typically incorporate several defining features that set them apart and contribute to their effectiveness: (i) Structure—these programs are often formally organized, characterized by a clear charter that outlines their mission and objectives, complemented by the appointment of an executive director, an advisory board consisting of expert members, and designated research or work groups specifically tailored to address distinct implementation purposes. This organizational structure includes objectives that focus on public outreach initiatives, effective communication strategies, information sharing, comprehensive fundraising efforts, human resource management, and the provision of essential technical expertise to practitioners in the field; (ii) Target—the primary focus of these programs may be directed toward the national government, addressing their pest management and information needs, the general public, specific industries related to agriculture, major pest groups that pose threats, and the development of best practices for pest control; (iii) Information—these programs actively generate a range of products grounded firmly in the practical experiences gathered from the field, complemented by relevant findings extracted from secondary literature concerning pest biology, ecology, and management strategies. This body of work also encompasses detailed papers and precise statistics quantifying the benefits and costs associated with IPM practices, as well as data on conventional pesticide usage patterns. Moreover, local expertise regarding a vast array of issues—including pest management practices, quarantine standards, market dynamics, rigorous cost-benefit analyses, and the environmental impacts—are frequently integrated into the program to ensure that it remains relevant and effective; (iv) A thorough and continuous examination of pest management techniques, predominantly focusing on IPM, is undertaken, alongside their foundational biological principles that support their effectiveness; and (v) Provision of general educational materials specifically aimed at the public which can potentially aid pest managers in their ongoing efforts, facilitating a better understanding of the essential concepts and practices of IPM that are critical for its success and sustainability.

IPM programs can offer crucial science-based and locally relevant information and skills that would otherwise be inaccessible to recipients. These programs also highlight the urgent necessity for effective crop pest management strategies that can adapt to the changing agricultural landscape. There is mounting demand for innovative crop pest control strategies in light of the failures of older methods, including the increased dependence on or development of new insecticides and fungicides, a ban on the use of methyl bromide fumigation, and a notable decline in the availability of agrochemicals for crop use, which has been largely due to a shift in the pesticide-testing ethos that prioritizes environmental safety and efficacy. A peer-reviewed biocontrol pipeline can effectively yield risk-free, planned releases of established controls and significantly enhance the global genetic understanding for crop pest management in that respect. Furthermore, there will be heightened expectations for transparency regarding the effectiveness of IPM programs, as well as accountability for consultancy practices and adherence to established standards. Recently, advanced biological resource discovery, propagation, and delivery methods have been developed, resulting in the emergence of bioherbicides, biopesticides, and various general biocontrol products. These innovations are intended for mass use in food production and have potential applications in fibers, textiles, and pharmaceuticals, which are all income-generating sectors that are crucial for sustainable industry growth and development.

8.4. Gene Drives for Invasive Species Control

Invasive species refer to naturalised populations of organisms that inflict, or are believed to inflict, significant economic or environmental harm (i.e., reducing agricultural production or leading to biodiversity loss). The long-standing geographic isolation of native species has fostered the evolution of life forms that are uniquely adapted to a specific region, particularly following the formation of various island ecosystems. These natural ecosystems, in turn, provide a multitude of essential services that hold great value for humans. Such services include fishing, hunting, recreation, and bio-prospecting for useful chemicals and resources that can be beneficial for medical and industrial purposes. The introduction of alien species into an island ecosystem—whether intentionally for agriculture or inadvertently via transport—has often culminated in the extinction of a considerable portion of the region's biodiversity. Such failures of island ecosystems to effectively compete against invasive species stand as some of the most critical examples of the ongoing biodiversity crisis that looms ominously over our planet today. However, it is crucial to understand that, although biological invasions may have reached their most visible and dramatic expression on islands, the reality is that they are not confined to these isolated landforms and can occur in a variety of terrestrial and aquatic environments across the globe. (Harvey-Samuel et al., 2017)

Invasive species are, by definition, a significant and pervasive trans-regional issue that impacts ecosystems worldwide. Many invasive species, whether intentionally introduced for agricultural purposes or accidentally brought into island ecosystems, are typically already present on a continental land mass or another nearby island. This previous presence provides a source population that continuously acts as a reservoir for invasions, effectively crossing geographical barriers with relative ease. For this reason, it becomes clear that ongoing and localized efforts aimed at controlling populations of individual invasive species will, inevitably, prove insufficient. These efforts will fail to effectively halt the continuing decline of biodiversity unless the source populations, which continuously replenish invasive species, are also systematically managed and controlled. The synergistic relationship between the introduction of invasive species and geographical isolation significantly exacerbates the biodiversity crisis, particularly in the period prior to the extensive creation of man-made barriers and land clearance for agriculture. Invasive species have benefitted from cultivated plants such as maize, bananas, and sweet potatoes, which have altered existing ecosystems to make them more susceptible to these invaders. This situation represents a critical and possibly final chance to prevent or decelerate the ongoing decline in biodiversity. Part of that comprehensive effort must encompass the armed surveillance of all arrivals to our ecosystems and the swift control of new pest species once they become established. Only through a concerted and multifaceted approach can we hope to protect the fragile balance of our ecosystems in the face of such challenges. (J. Mitchell & Bartsch, 2020)

Conclusion

The path to enhanced productivity is by ensuring that integrated pest management is executed regularly and effectively so that it is adopted by the greater part of the farmers in the region and agricultural sector. Reaching this goal entails overcoming social, economic, and sometimes legal obstacles. There is still a continuing lack of knowledge about crop protection, pests, and, in some instances, the required technology and installations are not available locally. Planning is needed at various political levels to encourage acceptance and practical execution. The arrival and spread of new invasive pests remains a threat to global agriculture (Pretty & Pervez Bharucha, 2015), this situation is only occasionally covered in the current pest management programs. It may be proposed that in these instances, measures need to be taken at the national or even international level to coordinate the acquisition of knowledge, technology, instruments, and products. Thought must be given to the fact that when new invasive pests arrive in an area, the existing pest management program may be totally useless. Hence, general planning is needed for these exceptional circumstances. Similar situations arise in development regions where both the crops grown and the pests may be totally unfamiliar, for example in the mountain regions of Central and South Africa. New regional pest management programs need to be designed from A to Z, including appropriate levels of knowledge and technology.

Locally adapted integrated pest management methods are available, but farmers are not sure about their appropriate application. In these cases advising methods need to be applied and knowledge transfer should be extended to institutions beyond the farm gate. In the old world countries who wish to farm sustainably, pest management needs to be more targeted, both spatially and temporally. It may be anticipated that targeted methods will also have better efficacy. This better targeting of pest management may be expected to be increasingly applied in the coming decades. Enhanced productivity may be by improving the effectiveness and sustainability of pest management and increasing the capacity for pest management research and development as a tool for development. Improved pest management has a gray side: it can be put to negative use and harm those involved, for example in the case of biopiracy. This growth has not always proceeded smoothly. There are opportunists and swimmers along the way. Generally, however, the large growth potential is expected to underutilized.

Sustainable agricultural intensification will depend on a proper balance of research, development, and education in the pest management system in relation to the local needs. A policy model has been developed for Integrated Pest Management. This model assumes that in order to keep environmental impacts of pest management under control, certain desirable land use changes are needed. Both crop and pest knowledge are important, which have to be taken into consideration in pest management assessment. Also, as in all decision making, normalization and weight need to be used to compare and aggregate different parameters.

Part III: Implementation & Case Studies

Chapter 9
Global Perspectives on IPM

Contents:

9.0. Global Perspectives on IPM

Integrated pest management (IPM) is a dynamic and multifaceted process of decision-making and action that is characterized by continual monitoring and consultation. This approach strives to effectively suppress pest populations while simultaneously minimizing the adverse impacts on human health and the environment. It serves as a viable alternative to the traditional approach of pesticide treatments, which heavily rely on the use of pesticides and often lead to significant environmental problems. The primary goal of IPM is to develop and implement a management strategy that maximizes crop production while also protecting the environment and promoting sustainability. Over the years, IPM approaches have undergone considerable changes and advancements. Initially, the focus was primarily on pest control technologies, which laid down solid foundations for effective pest management. However, there has been a gradual and noteworthy shift towards improving social, human, and natural capitals. This evolution seeks to address the needs of individuals and communities more effectively while simultaneously increasing the capacity and resilience of these essential capitals in sustainable agricultural practices. (Pretty & Pervez Bharucha, 2015).

Particular mention should be made of the Farmer Field Schools or FFS, designed as a means to increase understanding of the agroecosystem through discovery-based learning by observing, monitoring, and experimenting with a major component of IPM programs, knowledge-intensive strategies. Originally designed as a way to improve understanding on the ecology of the rice crop in selection with the brown planthopper, FFS for rice became an international success and platforms for imparting knowledge on a range of technical, economic, and social issues and an entry point to development processes. FFS has been established for diverse topics and crops ranging from crop and non-crop pest-management strategies, soil and water management practices, pest-disease crop biomass-biolitter management practices, and nutrition issues to postharvest practices in more than 90 countries. More than 10 million FFS participants have received training since 1994 and the number is increasing annually by 1 million. There have been matching changes in volume and venues. An increasing number of field schools are being organized by farmers themselves drawing upon a range of experts from universities, banking sectors, research, and extension. Also emphasized in recent years have been the placement of learning into practice, scaling-up and out programs, and gradual approach to tackle complex socio-economic or biophysical challenges.

9.1. Lessons from Successful Programs Around the World

By bringing together the diverse experiences drawn from a multitude of countries, regions, and localities, this document aims to set forth the most reliable and best available evidence pertinent to the pressing challenges of assuring food security while also promoting environmental sustainability. It draws heavily on the insightful lessons gleaned from successful programmes that have been implemented around the globe, and it also references new initiatives that have sought to adapt these valuable experiences to fit local conditions effectively. The overarching hope is that food producers situated worldwide can realize significant benefits from this rich material, allowing them to learn from it, thoughtfully adapt the insights based on their unique circumstances, and invest wisely in facilitating the necessary changes needed to ensure sustainability and productivity. As has been clearly evidenced through experience, the issues relating to pests, irrespective of their source, are mirrored across nearly every other crop group. As production systems become more intensive, the risks associated with various diseases are seen to increase correspondingly. However, these pressing challenges can be effectively addressed using low-cost, locally tailored practices that share remarkable similarities with the successful strategies documented in other crops, as elaborated upon in the subsequent chapter. The considerably wider geographic scope of the experiences that will follow promises to deliver invaluable insights into the larger landscape of how pest issues, which pose significant threats to the food security of impoverished communities, can be addressed effectively and sustainably. A multitude of organizations have undertaken the task of producing numerous "case studies" detailing their successes in this arena. The impact of these studies has been profound, as they have proven to be very effective tools in persuading managers, donors, and policymakers alike of the worthiness of these practices for investment. With each implementation of these practices, an increasing body of new evidence concerning their efficient effectiveness has emerged, enticing several stakeholders to believe that guidance on how to replicate and scale these successes was timely and necessary. After considerable deliberation regarding the enormity of the task ahead and how to effectively assimilate the wealth of considered thought and practice existing, it was ultimately concluded that through strategic utilization of the extensive previous materials already available, it would indeed be feasible to forge a highly sensible document focusing on experiences related to identifying where and how best to invest in transformative change aimed at food production and food security on a broad scale. The lessons learned from decades of experimentation in food production and pest management, disseminated and practiced across various continents, should be transferable to support efforts in addressing the current pest-related disasters that threaten sustainable production everywhere. Most governments and various agencies have, at some point, allocated funding towards research, trial initiatives, mass distribution of resources, and extension services across thousands of local sites, leading to the long-term implementation of beneficial practices in diverse communities worldwide. This collective experience has been an integral part of the challenge since the late 1980s when mass dissemination of locally relevant, cost-effective, and highly adapted solutions—such as the careful sourcing of seeds and the provision of vegetatively propagated planting materials—was first initiated. Several other crop groups have also confronted this challenge, and the knowledge and innovative ideas borne from these experiences should indeed be transferable across different agricultural sectors. This compendium of experiences, filled with valuable lessons learned and practical case studies showcasing successful models of food security, has been synthesised into a comprehensive "how-to" guide aimed specifically at enhancing food production capabilities in developing countries.

9.2. Community-Based Pest Management in Developing Regions

Integrated pest management (IPM) programs have now been in existence for approximately 50 years within developing regions of the world. Over this significant period, there have been numerous major achievements that have positively impacted agricultural practices and food production. However, there is an urgent need for a much greater focus on addressing the shortcomings that exist in the delivery of these programs. If the next 50 years are to witness a satisfactory increase in on-farm food production, alongside enhancing the industrial and service sectors, a concerted effort is required. Furthermore, alleviating poverty remains a critical goal, along with the provision of improved health and wellbeing for all segments of the population within these developing regions. The goal is to elevate living standards and ensure sustainable development across these areas. (Parsa et al., 2014).

A significant number of Integrated Pest Management (IPM) packages are currently available, and as far as the aspects of crops and pests are concerned, the essential technical foundation for managing technology and mechanisms for future developments is already established. However, exceedingly small areas are presently adopting IPM practices, and the over-reliance on traditional agriculture has unfortunately led to widespread poverty and the related social, economic, demographic, and environmental problems that arise from it. Institutions, encompassing both public and private entities, must urgently be strengthened in order to effectively accelerate the transfer of technology. To ensure that this transfer of technology is sustainable over time, it is crucial that there be a much greater emphasis placed on empowering farmers and promoting group action within the agricultural community. Additionally, the constraints and challenges related to technology transfer, particularly regarding the necessary on-farm and off-farm support systems, need to be thoroughly resolved to enhance overall effectiveness and implementation. (Pretty & Pervez Bharucha, 2015).

9.3. Collaboration Between Farmers, Researchers, and Policymakers

Over the past two decades, there has been a notable increase in the recognition of the significant necessity to engage effectively with farmers. Despite this progress, there continue to be persistent gaps that exist between the perceptions of researchers and the understanding of farmers regarding pest management practices. To address these gaps, fostering greater communication and dialogue between researchers and farmers is essential. Additionally, securing strong governmental support for Integrated Pest Management (IPM) practices can play a crucial role in bridging this divide. Furthermore, integrating pest management information into local language films can enhance accessibility and understanding among farmers. The involvement of farmer communicators can serve to facilitate the transfer of knowledge, while providing training that brings researchers and farmers closer together would also greatly improve the overall understanding of these practices. Ultimately, the active participation of farmers in IPM projects is not only beneficial but critical for ensuring the long-term sustainability of these initiatives. (Parsa et al., 2014).

Willingness to adopt Integrated Pest Management (IPM) is still significantly hampered by a variety of complex issues that are related to critical factors such as policy, sustainability concerns, and a lack of adequate support from NGOs, researchers, or service providers in the field. It is essential that the role of agricultural extensionists be increased and strengthened in order to ensure that effective IPM strategies are fully implemented in agricultural practices. Furthermore, there is an urgent need for enhanced support, collaboration, and communication between researchers and farmers regarding the dynamics of pests and pest management practices. This includes the timely dissemination of pertinent information between these two important groups to aid in coherent and effective decision-making. Priorities for future IPM research must include a thorough assessment of species damage through bold and innovative experimentation, as well as a deeper understanding of the diverse impacts pests have on crop yield, which ultimately influences the management strategies that farmers choose to adopt. Additionally, the investigation into the costs and benefits associated with labour-intensive pest management options is crucial, alongside efforts to encourage local adaptation and comprehensive evaluation of effective control options against pests.

Learning is indeed a two-way process, but many experiences suggest that awareness and appreciation of various pest management options, along with understanding farmer perspectives outside of the scientific community, is still a relatively new phenomenon. There is a pressing need for better understanding within the research community regarding how to effectively improve mechanisms that facilitate better knowledge flow and enhance decision making at the ground level. Several successful initiatives have been undertaken to foster collaboration between scientists and farmers, but there has been insufficient attention on the lessons learned from these valuable initiatives. It is crucial to analyze and document what has worked and what has not, in order to develop more effective strategies moving forward. Additionally, there is a significant need to broaden the ongoing debate on pest management to comprehensively evaluate farmers' unique knowledge, adaptability, and ethical considerations in effective pest management systems. Engaging farmers in these discussions could lead to more sustainable and locally adapted pest management practices.

Chapter 10
IPM Across Different Cropping Systems

Contents:

10.0. Introduction

Integrated Pest Management (IPM), which was originally introduced in the 1960s for the focused control of rice stem borers affecting crops in the Philippines, has since evolved into a widely applied, ecologically based pest management approach that is utilized in various agronomic crops around the globe. This approach emphasizes the importance of coordinating a diverse array of pest control strategies and actions that are tailored to the specific conditions of the crops being cultivated, the environmental context, and the types of agricultural inputs that are employed. It involves an assortment of methods, including cultural practices, mechanical interventions, trapping techniques, and pheromone applications, all of which are critical for effective pest management. When it comes to rice cultivation, various cultural techniques have been identified as being particularly effective in controlling pests. These techniques include the use of stale seedbeds, efficient water management systems, the selection of different crop varieties, implementing crop rotation strategies, and utilizing nitrogen fertilization, among others. However, despite its benefits, there are several limitations of IPM when applied under actual field conditions. These include a heavy dependence on a limited number of cultural and mechanical practices, difficulties in predicting pest resurgence at later stages, and instances where pest shifts have led to failures in recommendations for control measures. Moreover, IPM strategies have also been developed for managing pests in fruit crops, such as mango, guava, lychee, pomegranate, banana, and a variety of vegetables, utilizing techniques such as variety screening, pheromones, and traps. In Maharashtra state of India, an implementation of IPM for grapevines has shown remarkable success, noting a significant reduction in the usage of pesticides. Additionally, advancements in technology have led to the standardization of prediction models based on Geographic Information Systems (GIS) and Remote Sensing (RS), which have been specifically designed for the effective IPM of grassy and broadleaf weeds in sugarcane farming.

Globally, there has been a significant prioritization of bio-inputs, biological control agents, traps, and newer, more environmentally friendly pesticides for their inclusion in Integrated Pest Management (IPM) strategies. The value chain analysis of various cropping systems for assessing the impact of IPM has been initiated and scaled up in the Eastern Indo Gangetic Plains region. Insecticides that fall under derogatory clauses require mandatory applicator training, thorough residue analysis, and effective farmer awareness programs to ensure safe usage. Recently, there has been a renewed emphasis on the implementation of IPM methodologies, particularly focusing on increasing the scale of operations to enhance efficacy and effectiveness at the state level. Currently, IPM is being put into practice through various platforms, including Common Service Centers, the National Agriculture Market, and Integrated Command and Control Centers. These initiatives are aimed at improving the accessibility of vital technical information to farmers and stakeholders while also working to rationalize prices in the agricultural sector. (Pretty & Pervez Bharucha, 2015).

Structural adjustment policies have notably reoriented the roles of the state, shifting their focus from being predominantly production-oriented to embracing marketing strategies, financial interventions, and business facilitation across all agricultural activities. This significant transformation resulted in a substantial reduction of state involvement in professional extension services and the delivery of essential agro-inputs in agriculture. Consequently, agro-input markets transitioned into a stage of complete privatization and lacked regulation, which had adverse ramifications for smallholder farming. This shift led to a concerning monopolization of resources and exploitative pricing practices that harmed the very farmers who were meant to thrive. As large-scale mechanized farming dominated the agricultural landscape, accompanied by intensive and pervasive use of chemical fertilizers and pesticides, agro-inputs in India garnered a notorious reputation as mass-poisoners of the land, contributing to environmental degradation. Signs of growing awareness regarding escalating income inequalities and the social polarization emerging within farming communities became more pronounced, reflecting deepening agrarian distress. Alarmingly high rates of farmworker suicides, negative repercussions of climate change, and extensive migration patterns added to the complexity of the issues at hand. Furthermore, the principles of food sovereignty began to feature increasingly in dialogues surrounding the intricate connections between agriculture and the environment, raising critical questions and urging for a more equitable and sustainable approach to agricultural policy.

10.1. Field Crops (Corn, Rice, Wheat – Focus on Global Examples)

The single most economically significant crop globally, and the most critical grain crop cultivated throughout the Americas, is undoubtedly corn. Moreover, transgenic corn engineered to contain a Cry protein sourced from Bacillus thuringiensis (commonly referred to as Bt corn) has seen widespread adoption among farmers and has effectively delayed resistance in the fall armyworm, a prominent pest. Interestingly, the same protein present in corn has also been utilized by a commercial entity operating in natural habitats and surrounding agricultural areas, specifically targeting the delay of resistance in the maize weevil in Colombia. The efforts to slow down resistance related to new Bt rice have significantly expanded the acceptance and popularity of Integrated Pest Management (IPM) strategies within the global rice market, ultimately saving multiple millions of dollars. On another front, the AquaAdvantage transgenic fish has faced delays in its introduction to the market primarily due to growing concerns regarding its potential impact on wild-type salmon populations. Notably, some major case studies have been conducted to delve deeper into these impacts and provide valuable insights. Additionally, the most successful perennial grain developed to this day is a unique dwarf hybrid of wheat which has gained attention for its agricultural benefits. Moreover, around three hundred crop species, many of which are indigenous to regions such as Africa, Asia, and the Americas, have been gathered and assessed with the aim of integrating them into smallholder farming systems. Examples of globally sourced crops that have been integrated and increased in traditional contexts include amaranth, quinoa, buckwheat, and itch grass. Furthermore, crops native to North America, which possess the remarkable ability to potentially double the yields of corn, sorghum, and soybeans, were carefully selected as structural models to replicate. It is crucial to understand that agricultural development is significantly influenced by various nonapparent processes. For instance, threats posed by diseases can reduce overall farming productivity and drive a shift towards subsistence farming; however, the often-invisible demand for multi-crop disease resistance plays a vital role in enhancing food security rather than simply diminishing it. Experiments related to the On farms soil disease hypothesis have uncovered the presence of soil-borne viruses harbored in the earth, which dramatically increased productivity trends through the cultivation of perennial crops that exhibit fast growth, drought tolerance, and resistance to diseases such as sorghum and others. (A. ANDERSON et al., 2020)(Pretty & Pervez Bharucha, 2015).

10.2. Horticulture (Fruits, Vegetables, Greenhouse IPM)

Though significant improvements in pest management have been achieved over the past decade, including advanced forecasts of pest occurrence and their phenology, there remain numerous open questions and remedies that need to be addressed and adopted effectively. For instance, there has been a somewhat thoughtless examination of climatic factors and their effects on various pests and their antagonists, which calls for a more thorough investigation. Furthermore, the feedback mechanisms between meteorological algae and pests need deeper exploration, especially with current analysis methods relying heavily on report-every-five-days auto-similar techniques. Pest forecasting at an extremely fine spatial resolution is still required to enhance our understanding and response strategies. It is critical to develop a better description of phenology segmentation that aligns closely with essential climate parameters. Moreover, it is necessary to push for further innovations in compatible low-drift spraying equipment. This includes designs such as booms equipped with see-through openings to enhance operator visibility, as well as the innovative design of surface-reducing emitters that minimize drift during application. There is also a vital need for refining the spray droplet extraction procedure to improve accuracy and effectiveness. On the theoretical side, modeling of escapes, tumbling spray, and diffusion processes can be adeptly described using Langevin equations, which can lead to better optimization of the timing of pesticide applications. Targeted bioremediation of pesticides using specialized microorganisms, as well as utilizing pheromones in conjunction with xenobiotic degrading enzymes, is an area ripe for further exploration. A crucial aspect is pheromone capture, especially in relation to the phenology of both the pest life stages and their antagonists, to ensure effective pest regulation. In addition, research efforts are directed towards identifying and testing acceptable alternative active ingredients for pesticides that could prove to be less harmful. Although certain compounds have been excluded from new registrations based on EU criteria, there is potential for their adoption under special procedures which could aid in pest management efforts. Moreover, law enforcement remains vigilant, ensuring close scrutiny of misapplications, such as the inappropriate use of pesticides on ambulances, since consumers today are increasingly aware of agricultural practices and control protocols that extend beyond Integrated Pest Management (IPM) strategies. Overall, there is a multitude of approaches in pest management that need to be analyzed collectively, as they may function effectively for the same pest species. By aggregating these various strategies and findings using collision diagrams, we can foster a more holistic understanding and better control mechanisms in pest management. (Damos et al., 2015).

10.3. Stored Grains & Post-Harvest Pest Management

Global postharvest losses of key cereal crops, beans, and a varying array of vegetable commodities are currently estimated at a staggering and alarming total of approximately 67 million tons in developed regions. The situation is even more concerning when we look at developing countries, where the estimated losses reach a terribly high level of 151 million tons. A notable and significant portion of these losses—approximately 49%—can be directly traced back to a multitude of different forms of pest infestations that commonly pervade storage areas of harvested grains and vegetables. Due to their diminutive sizes and tightly knit life cycles, a wide array of diverse arthropod pests severely infests stored agricultural commodities. These pests encompass not only grain weevils but also flat grain beetles, grain moths, various types of bristletails, and several topweevils. Additionally, the natural adversaries of these pests, including foil wasps, dung beetles, and even spiders, play a complex role in the dynamics of pest management. Among smallholder farmers, chemical dusts such as Pirimiphos-methyl have become increasingly prevalent and common as the method of choice for tackling these pervasive postharvest pests that threaten their yields. However, the consistent reliance on low Integrated Pest Management (IPM) approaches strongly correlates with suboptimal pest management efficiency, and it often results in compromised produce quality. These challenges stem from serious infestation issues that frequently go unaddressed, leading to further complications in the postharvest phase. The trading systems for smallholder maize operate in a manner that regrettably facilitates an open cycle for the movement of pests from poorly managed stockpiles to those that are better maintained, thus perpetuating the cycle of infestation that leads to further losses. Moreover, the multitude of neglected and disused storage facilities also serves as a prime habitat for these pests, thus exacerbating the already daunting boundaries of postharvest infestation challenges. Maize grain itself acts as a conducive reservoir, supporting both the persistence and movement of pests across various agricultural landscapes, which leads to grave concerns regarding food security and quality. These observations strongly indicate a substantial pest management dilemma, existing both within and among storage facilities, while exposing numerous opportunities for tiny pests to exploit vulnerabilities that are present in these systems. To date, integrated pest management programs have diligently delved into a broad range of genetically and environmentally sustainable alternatives aimed at significantly reducing the losses incurred due to invasive grain pests after the harvesting process has taken place. This meticulous technical approach conscientiously takes into consideration various factors such as taxonomy, pest behavior, ecological impacts, biological characteristics, and the management tools applicable to insect pests in the field. Furthermore, it accounts for the vast variety of agro-ecological conditions and specific management scenarios prevalent in developing nations, their impacts throughout pest development, and current practices in use. The programs suggest improvements to existing practices, while also recognizing knowledge gaps that linger and need to be addressed in pest management strategies. Hermetic storage structures have been predominantly deployed and thoroughly tested against a variety of postharvest pests. This includes several that have been previously identified as concern pests like bruchids and various types of beetles that can wreak havoc on stored grain. For instance, metal silos with a capacity of two to three tonnes at the farm level, alongside portable hermetic bags that can hold up to 50 kilograms, have been successfully utilized by farmers in Kenya. These innovations aid in the successful marketing of their agricultural produce, particularly through the usage of 70 kg tractor-transportable jute or hemp bags, which have shown to be effective in significantly reducing the losses caused specifically by Sitophilus zeamais. These innovative hermetic storage materials provide a crucial opportunity to mitigate significant pest losses for smallholder farmers, or for those operating within storage cooperatives, thereby enhancing and improving the overall postharvest management process in a meaningful and impactful way. This approach not only assists in reducing waste but also opens doors for improved food security and sustainability in agricultural practices. (López-Castillo et al., 2018; Talwana et al., 2018).

10.4. India's Non-Pesticidal Management (NPM) in Cotton

Agriculture is the most important pillar of livelihood for the people of India. The Indian farming community has been facing increasing problems in terms of production and marketing. The problems of production include the production of safe food, unregulated exploitation of chemical pesticides, health hazards to farmers due to the toxic nature of the pesticides, pesticide resistance in pest species, hygienic and industrial processing of cotton, and other fiber crops. The implementation of genetically modified cotton has further complicated these problems. This has necessitated a halt in the further advent of synthetic pesticides and advocates for alternative non-pesticide pest control measures. In this context, non-pesticidal management (NPM) of pest control is discussed here.

Russell et al. (2000) discussed the basis for the development of NPM. In order to derive concepts for the development of a NPM strategy, the overall pest control process is analyzed in terms of the mechanistic 'schema' impacting the cash crop raw output to reveal the functions and parameters of the agro-ecosystem. Thus, the control measures, both preventive and corrective in nature, which can be used for any crop-pest combination, were identified. A simple representation of the agro-ecosystem was generated to highlight the principles to be used for the management of the agro-ecosystem and its aggressors. The principles are to enhance naturally occurring checks of erstwhile harmful agencies. The checks considered for treatment include the action of non-foliar feeding insects, the activity of natural enemies, the use of resistant varieties, cultural means, and the reduction in chemical pesticides. The principles were opted with a view to use them in different NPM models depending upon their environmental/social acceptability and their incremental cost compared to IPM for large scale application.

NPM models were established and thoroughly tested on farmers' fields across various regions in India, successfully covering an impressive area of more than 22,900 hectares. In conjunction with these field trials, a significant number of benefit-cost analyses were conducted to evaluate the effectiveness of the models. Remarkably, in nearly all of the trials, herbivore pest populations remained below the critical economic injury and yield loss thresholds. This successful outcome was largely attributable to the activities of the checks that were highlighted and emphasized within the models utilized. Following these promising results, a collaborative joint activity was launched at a national level, focusing on cotton-based NPM systems with the objective of deriving an 'India-wide NPM strategy.' This initiative was seen as a forward-thinking approach, aimed at ensuring the long-term sustainability of cotton productivity, while maintaining effective pest control measures without relying on the use of harmful pesticides.

Chapter 11

Scaling Up Integrated Pest Management: Policies and Farmer Adoption

Contents:

11.0. Introduction

Current climate change is having significant effects on food systems. Crop yield responses to temperature and precipitation extremes are widespread, with a stronger warming effect in staple crops. However, yield declines were also found across developing and high-income regions for both hot days and nights. This can increase food prices by 180% across the globe, increasing malnutrition by 193 million people. Since climate change is a global problem often requiring local solutions, countries are expected to vary in climate change exposure. The effects of climate change on countries across the world are substantial but poorly understood. This information is critical for governments and international organizations to increase resilience. Agricultural production strongly ties to weather fluctuations, making it sensitive to climate variability and extreme climate change. The global mean temperature was estimated to increase by 2.1 to

3.9 degrees Celsius over this century, and inequalities in daily temperature and precipitation extremes would widen as countries become more warmed. Current climate change can thus have considerable and various effects on global food systems. One proposal is that global gains in yield could not offset losses in key regions, especially for wheat, underscoring the threat of climate change to food security. Stimulation outputs explore how cereal yield changes from 1981 to 2010 downscaled to grid resolution in 2.5 against future climate change. With the expectation of continued climate change, most rice and corn-growing areas are likely to see yield increases, while wheat yield declines dominate in Pakistan and Central Asia, leading to altered cropping areas. Adoption of agricultural technologies, especially those developed for sustainability, is critical for sustainability transitions. Smallholder farmers hold important potential for adopting the technologies needed to address these issues, but innovations struggled to achieve widespread adoption (Pretty & Pervez Bharucha, 2015). For agricultural sustainability, farmer organizations or a multi-faceted approach could be effective in scaling-up standards or policies. Questions of technologies, insights on how farmers came to adopt the technologies, how knowledge on the technologies was introduced, and what policy measures helped farmer adoption are central issues of the recommendation.

11.1. Government & NGO-Led IPM Programs

Governments and NGOs have made significant efforts to address and overcome the various barriers that hinder farmer adoption of Integrated Pest Management (IPM) through a range of large-scale training programs. These initiatives are typically structured to include the training of trainers—especially focusing on agricultural extension staff—which is then followed up by comprehensive training sessions designed specifically for farmers. Generally, there are two primary types of programs available: large-scale national programs that are initiated by government entities and prominent international NGOs, as well as local programs that are undertaken by regional and national NGOs that function at a more localized level. Prior to the introduction of such programs, agricultural extension agents primarily relied on delivering technical messages to farmers on an individual basis. This approach heavily depended on the use of printed materials, which often limited the effectiveness of the communication process. Farmers had minimal opportunities to actively participate in these training sessions, often remaining as passive listeners to the instructions provided. For instance, an IPM training program that was orchestrated by a local NGO during this period usually consisted of a 2-day training event, where only a few hundred farmers would attend. This approach primarily included lectures that did not encourage interaction or engagement. Consequently, many farmers were not particularly inclined to read the printed materials supplied to them and developed a skepticism towards the contents provided, primarily because the concepts and techniques presented appeared too foreign and disconnected from their practical experiences. As a result, they remained practically unchanged in their long-standing pest management strategies and practices. However, soon after the commencement of the aforementioned sustainable agriculture initiatives, there was a transformative shift, marked by major investments directed towards farmers' training programs. IPM was increasingly recognized as a crucial and integral component of integrated farming systems that are essential for modern agriculture. This approach was specifically aimed at enhancing agricultural productivity, fostering economic development, and promoting social equity in farming communities. IPM was envisioned to serve as a vital conduit to optimal land-use practices, sustainable resource management, and a significant increase in biodiversity across various agricultural landscapes. In the Philippines, for example, the economic impact of rice crop losses due to planthopper infestations has been alarmingly estimated at around 1 billion pesos each year, which is approximately equivalent to 20 million USD. In response to this dire situation, the national government took decisive action by implementing a ban on the importation of chemical insecticides. However, this well-intentioned policy soon faced challenges as an influx of illegally imported chemicals began to flood the market, undermining the effects of government regulation and highlighting the complexities involved in managing agricultural pests in a rapidly changing environment. (Pretty & Pervez Bharucha, 2015).

11.2. Farmer Field Schools (FFS) & Extension Services

The Farmer Field Schools (FFS) extension method was innovatively established in Southeast Asia back in 1989, with the primary goal of training farmers on the effective application of Integrated Pest Management (IPM) programs. This initiative specifically aimed to encourage a significant transition away from the reliance on synthetic pesticide use and promote instead more environmentally friendly, socially responsible, and economically sustainable pest management practices and strategies. The FFSs are organized as season-long participatory learning programs, which meticulously instruct farmers in the nuanced aspects of pest management grounded in the principles of agro-ecosystem analysis. Notably, the Farmers Field Schools (FFS) exemplify a widely adopted participatory approach to agricultural extension, benefiting from a rich and extensive history within the realm of agricultural development and support. By engaging in the FFS, farmers are empowered to critically analyze their fields as dynamic agro-ecosystems through systematic pest and disease observation, comprehensive life cycle studies, in-depth field asset mapping, and thorough production alongside pest risk factor analyses. The training provided significantly enhances the participants' capabilities in assessing pest and disease levels present in their fields, while also considering crucial aspects such as gender and age dynamics, various income-generating strategies, and important labor considerations that impact their farming practices and livelihoods. This holistic approach fosters a deeper understanding of the interconnectedness of these factors, enabling farmers to make informed decisions conducive to sustainable agricultural practices. (Mariyono et al., 2013).

Over the years, the FFS approach has undergone significant transformation and evolved into a multitude of variants, which include but are not limited to, crop-specific Farmer Field Schools, eco-FFS, bio-FFS, and Health FFS, just to mention a few notable examples. This paper presents findings from ongoing research that assesses the effectiveness of the FFS approach as a valuable tool in promoting integrated pest management (IPM) programs. The research revealed that higher levels of pest control knowledge and a noticeable decrease in pest damages were positively associated with active farmer participation in FFS initiatives. Nevertheless, there is a pressing need for a robust long-term support mechanism that can effectively sustain the adoption of these technologies over time. The primary objective of the study was to thoroughly describe the effectiveness of the FFS extension method, focusing specifically on its significant impact in encouraging the adoption of integrated pest management programs among farmers. Additionally, the research identifies critical success factors that are essential for promoting the widespread adoption of the FFS method, while also addressing the various challenges that participants may encounter along the way. (Akuno et al., 2014).

11.3. Economic Barriers to IPM Adoption

The adoption of holistic pest management approaches in agricultural settings faces several constraints that hinder its widespread implementation. These limitations are rooted not only in the gaps in knowledge and information that farmers possess about holistic pest management but also in the economic risks that are tied to that knowledge. Despite a general tendency toward risk aversion among farmers, studies have revealed that they can still show a willingness to embrace new knowledge and product offerings, particularly those that are introduced by entrepreneurs who are more inclined to take risks. Therefore, for holistic pest management practices to gain traction, an improvement in the understanding of pest biology, ecology, and management strategies is crucial. To facilitate this shift, new policies aimed at mitigating the existing risks and uncertainties surrounding pest management options could be introduced. These policies could focus on transferring some of the financial risk that farmers bear associated with pest management to the financial sector, creating safer avenues for farmers to explore holistic practices. For example, developing "insurance companies" that operate on various scales could play a significant role in pooling the economic risks that farmers face due to pests. By offering targeted pest management products and services that are transparently marketed and communicated in a credible manner, these companies could enhance farmers' confidence in adopting holistic approaches. This strategy would be particularly advantageous in situations where multiple farmers experience similar pest challenges that can be anticipated regarding their severity, timing, and spatial occurrence, for instance, with the entry of a pest species into novel geographical areas. However, implementing such a model of risk sharing would require substantial investments in financial, human, and social capital, which, in turn, could prolong the time frame for full realization of these benefits. Moreover, for these comprehensive pest management strategies to gain widespread acceptance, clear and tangible economic incentives must be evident for farmers currently engaged in both holistic and conventional pest management practices. The economic benefits could encompass aspects like increased crop yields, decreased pesticide expenditures, and even lower labor costs or reduced expenses linked to extra crop-caretaking efforts. To substantiate these economic claims and support the proposed models, rigorous research is essential to bridge existing data gaps regarding the impacts of various pest management strategies and their intricate interactions. This may also necessitate longitudinal studies designed to dissect the underlying reasons why certain pest management approaches have found traction in some regions while remaining less popular elsewhere. Furthermore, it is vital that the products and services created through this transformation are also unobtrusive, ensuring that they do not mandate actions that the farming community does not see as justified or worthwhile. Public policies could greatly assist in transforming pest management practices by framing the desirability of adoption not merely from an abstract notion of 'Rationality,' but instead on the practical understanding that the current system of pest management is inadequate. Additionally, frameworks could be established that avoid expressing adoption thresholds in percentage terms, as this might imply a misrepresentation of the actual adoption rates of particular pest management methods. Instead, determining these thresholds could focus on the likelihood that all farmers would adopt a specific pest management strategy if such an approach were made viable and beneficial for them. (Pretty & Pervez Bharucha, 2015).

11.3.1. Cost Analysis of IPM Practices

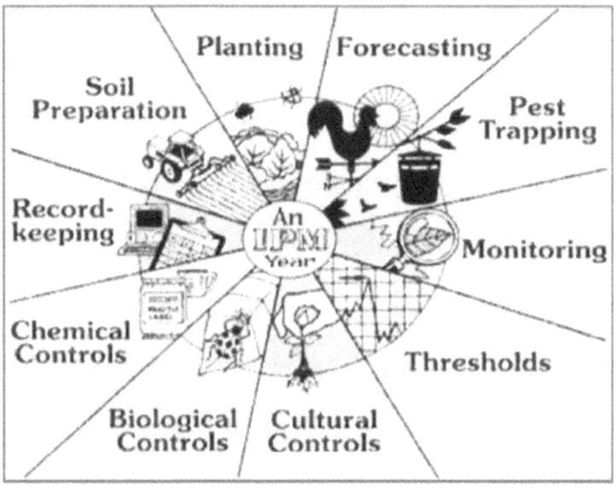

An extensive analysis of the various costs and benefits linked to the differing adoption rates of Integrated Pest Management (IPM) practices among farmers is presented here, based on the comprehensive data that has been meticulously collected. In order to acquire the necessary data for this analysis, farmer surveys were conducted alongside an examination of prevailing market prices. The relevant costs associated with IPM were estimated on a per-hectare basis, ensuring that farmers could readily comprehend the financial implications of each practice. Notably, most of the costs associated with implementing IPM were observed to be quite similar for both practices utilized across the three countries investigated. When it comes to the benefits derived from IPM, increased yields obtained in the various field trials were computed as group averages. However, it is crucial to note that in several instances, farmers themselves experienced even higher levels of benefits in the practice plots that were not considered within the initial scope of this analysis. For instance, farmers applying specific cotton practices in Khamti achieved a remarkable yield benefit of 524 kg/ha in comparison to the control plots, highlighting the strong benefits derived from farmers' own practices. By educating farmers on how to utilize cost-effective methodologies, there was a significant reduction in pest infestations, which in turn led to increased marketable yields and a substantial boost in overall farm income. The implementation of IPM strategies yielded a net benefit that varied considerably between farmers in China and Uganda. In many cases, the costs associated with the majority of IPM practices were either negligible or minimal. Specifically, in Uganda, even when accounting for the complexities of confusion pricing, IPM practices successfully generated an impressive $1,072 per hectare, culminating in a remarkable 97% return on investment. Conversely, for farmers in China, although there was the inclusion of a 5% case-sharing amount designated as costs, the yield benefits remained significantly substantial. Most of the costs associated with the eight distinct IPM practices primarily revolved around inputs, rent, inherent risks, and the ongoing monitoring of market prices. Furthermore, for the newly reinforced aversion practice concerning spraying only when necessary—under the IPM strategy in Bangale—over the year following the case-sharing initiative, farmers witnessed a staggering benefit amounting to $258,919,807, which equated to a 70% return on investment. This was largely due to the enhanced practices that went beyond merely utilizing yellow sticky traps on ethnic cash crops, implemented in a timely manner to mitigate significant harm and maximize efficiency.

Sustained yields for as long as nine years while utilizing integrated pest management (IPM) practices, all without experiencing an increase in farm input costs, were clearly indicated through the survey results. The majority of the surveyed farmers who had been implementing IPM practices did not choose to adopt any additional high-cost best practices. This suggests a significant trend among farmers towards a more cautious approach when it comes to investing in costly interventions. Furthermore, when a farmer began employing a better agricultural practice, he often encountered challenges in accurately estimating cost reductions and weighing the decision on whether to continue utilizing it. These findings highlight a significant gap in awareness concerning the costs associated with new practices, while there is a clearer understanding of the realized gains from implementing IPM methods. A crucial question pertaining to the sustainability of IPM interventions revolves around the costs involved, as well as the long-term impact on the knowledge possessed by farmers. The data collected showed that an impressive 82% of farmers in Uganda were capable of naming, assessing, and modeling the behavior of both large and small flea beetles. This level of knowledge is commendable, especially when compared to previous analyses that suggested a failure to sustain increased agricultural yields after two to three years following the initial intervention. The insights gathered from this survey may serve as a valuable foundation for further investigations into effective and sustainable agricultural practices within the context of IPM, particularly how these practices can lead to lasting benefits for farmers in the long run.

11.3.2. Market Incentives for IPM

Market incentives for integrated pest management (IPM) have significant potential to overlap with the funding frameworks established for agri-environmental policies (AEP). It may indeed be feasible to deliver relevant IPM products through the channels provided by AEP programs. In order for these financial incentives to effectively encourage the uptake of IPM practices across substantial areas of farming land, it is critical that they be robust and significant enough to motivate farmers. Furthermore, measures must be implemented to ensure that farmers not only adopt IPM but also continue to adhere to those practices over time. Alongside these financial enticements, considerable efforts must also be directed toward ensuring that both AEP and IPM frameworks are rolled out in an effective and efficient manner. The willingness of farmers to comply with environmental regulations is a prerequisite that allows for the successful use of market incentives in this sector. The reform legislation under the Common Agricultural Policy (CAP) that seeks competitive bid prices can contribute by offering insightful information on what farmers may require financially, substantiated with evidence, particularly in relation to developing and implementing IPM strategies. In addition to direct market incentives, evidence suggests that it is unlikely that national regulations will serve as a sufficiently strong catalyst to lead farmers to adopt IPM practices across extensive tracts of land. Research indicates that a combination of rigorous public policy support, when paired with proactive private sector participation, tends to yield superior outcomes for IPM adoption. The current legislative landscape regarding pesticide use is in a state of flux, and the process of reforming regulations in this domain presents numerous challenges, as it often necessitates a considerable shift in perspective about the associated risks and benefits of pesticide use. Since the year 2001, legislative frameworks established by the European Union have significantly propelled awareness of IPM among farmers, as they mandate that each farm must have a comprehensive IPM plan implemented. These plans can be effectively delivered using self-audit questionnaires, which have demonstrated efficacy, especially when compared to more complex and cumbersome planning frameworks that may deter compliance. Recent initiatives undertaken in England have showcased that when such programs are combined with some degree of public policy funding, the investment of time, and the use of either straightforward products or those integrated into existing marital processes, it can greatly facilitate widespread coverage of IPM practices. Standardized questionnaires for IPM audits that are user-friendly may serve as a viable means of addressing various discrepancies in farmers' levels of understanding regarding what exactly is expected from these plans. Reports indicate that compliance rates were similar between farmers participating in Countryside Stewardship Management Plans (CPMP) and those engaged in Environmental Land Schemes (ELS), particularly when focusing on a comparison of these two policy approaches. Without the presence of publicly funded market incentives, farmers who prioritize stable and consistent policies represent the most appropriate target demographic for future initiatives aimed at encouraging further uptake of IPM practices. (Pretty & Pervez Bharucha, 2015).

11.4. Regulatory Frameworks for IPM

Among the many tactics and management practices available to growers to control crop damage inflicted by pests is integrated pest management (IPM). In its broadest definition, IPM is the weighing of all the available information on crop alternatives, pest biology and behavior, and the proper timing, choice and application of pest control tactics to achieve cost-effective management while minimizing adverse environmental and human health effects. Under fundamental IPM principles, meaningful options are sought to: 1) prevent pests from reaching economically damaging densities; 2) nullify pests' growth potential to rebuild their populations; and 3) use the best methods available to keep damaging insect pest densities below these levels afterward. Environmental agencies have developed regulatory frameworks to carry out mandates that have both helped and hindered efforts to find and implement effective pest control tools that meet IPM criteria.

Pesticide regulation, including the registration of biopesticides that would more easily meet IPM goals than traditional chemical pesticides, is perhaps the most significant way that government regulation can assist efforts to integrate biological control methods into existing pest management programs. Environmental agencies have developed such mechanisms in many countries to expedite the approval of environmentally safer pest control alternatives to conventional chemical pesticides. In the USA, for example, Congress passed the Pesticide Registration Improvement Act of 2003. Under its provisions, the USEPA is to promote shorter decision review periods for applications for reduced-risk, biopesticide and public health pesticide applications for at least four years after adopting regulations to implement the Act. In addition, the Federal Food, Drug, and Cosmetic Act was amended in 1996 to include the Food Quality Protection Act (FQPA) which required the USEPA to re-evaluate all existing pesticide tolerances to ensure they met current health-based safety standards. The FQPA also mandated the establishment of a registration review process that would require every pesticide registration to be reassessed every 15 years to consider new safety information and determine whether to amend, renew, revoke or continue the registration under the Federal Insecticide, Fungicide, and Rodenticide Act (FIFRA). These initiatives will provide IPM practitioners and growers with new and safer tools for IPM programs (R. Matten et al., 2008).

11.4.1. Overview of Quarantine Regulations

Millions of tons of plants, plant products, and other regulated articles move across international borders each year. In many countries phytosanitary legislation and regulations govern the importation of plants for planting. Whereas there is tremendous variation in approach and interpretation of phytosanitary matters, typically a national plant protection organization (NPPO) is empowered by legislation. Separate laws cover what can be imported, and regulations specify how these imports are handled. Many countries have adopted the International Standards for Phytosanitary Measures (ISPMs) by which they are expected to implement effective pest risk analysis (PRA) systems to support rational regulation and management of trade and native agricultural and landscape plants (Soares dos Anjos Marques et al., 2016).

The theme, specifically Section 1, presents the fundamentals and a number of biological and technological innovations that enable the identification of new pest threats. These contribute to many of the new methodologies developed in Australia, the US, NZ, and South Africa, which significantly enhance pest detection and risk management. Section 2 highlights significant biosecurity threats and pest issues that challenge the agricultural and horticultural industries, such as pests on the list of exclusions/unwanted more advanced tools than those presently used in countries such as the EU and the US. Section 3 outlines legislation and regulatory systems handling risk items once onshore.

Most countries continue to face a range of complex and interrelated biological, biophysical, environmental, socio-economic, and regulatory issues, yet increasingly have fewer resources, people, and dollars to manage plant health. Improved monitoring and risk reduction practices and technologies that can target susceptible plants /pathways, improve Public awareness and compliance, and advance pest-free place of production systems will benefit pest management on a broad scale. Quarantine pest issues, chemical use, systems approaches, and advances in training and skills are addressed. When engaging with an audience of diverse expertise and knowledge, pooling information in advance can catalyze discussion and promote a common understanding of issues to be addressed (Alexandre Nogueira de Sá et al., 2016).

11.4.2. Definition and Purpose of Quarantine

Quarantine is established by law within a country or geographical area to erradicate insects, fungi, bacteria, nematodes, weeds, etc., that are dangerous to agriculture and that can be introduced and spread through infected food and plants (Soares dos Anjos Marques et al., 2016). Information on pest distribution and potentialities of organisms regarding import quarantine agencies is important for the assessment of potential negative interactions between pests and their new environments leading to destructive invasions, thus to help in the choice of suitable resistance crops and lists of organisms. Quarantine regulations vary among countries. The analysis of infection risks must be country-specific, pest-specific, and product-specific. Inspection and monitoring activities are of key importance in pest quarantine systems. A pest is subjected to quarantine whenever it is prohibited to transport it across international borders, or whenever its internal movement is restricted between natural zones or regions within a country. The fields of action of a quarantine regulation scheme include: a) Legislation; b) Risk analysis; c) Import inspections; d) Monitoring and pest management; e) Good practices; f) Surveillance; g) Universities, extension, and research centers; h) Preparation of quarantine projects. The official actions act in a preventive way which allows reducing risks prior to the introduction of potential external threats, but do not act with corrective or control measures regarding internal plant pest populations.

11.4.3. Types of Quarantine Measures

Quarantine measures are actions taken to prevent pest entry, establishment, or spread, focusing on pathways with significant pest risk. Quarantine measures include tabled (or lender) and operational measures. These can be legislative or technical measures. Legislation and legislation complement each other: the first defines an economic and biological context, goals, and responsibilities, while the latter decides on operational aspects such as timing, location, and resources. Operational measures can be official or non-official and generic or specific. They involve direct or indirect pest detection and defined control measures. Some non-official and generic operational measures are hidden from potential passengers or shippers. Simple examples include inspections at sea in favourable conditions to allow joists to be cleaned (Soares dos Anjos Marques et al., 2016). As invisible to the shipper as reptiles or taints had previously crossed the border at hidden locations, it is impossible to check without active technology en masse. However, on occasions, these measures are not enough to keep pests outside their target territory. In these situations, specific and official biosecurity measures are appropriate.

Specific and official quarantine measures are actions taken to prevent an outbreak when pest establishment is suspected. They cannot be carried out on a quarantine zone (QZ) but need a necessary quarantine area (QZ). Quarantine areas can be habitats where the pest is suspected of being present or habitats known to be infested.

11.4.4. EU's Sustainable Use Directive (SUD) on Pesticides

The European Union's Sustainable Use Directive (SUD) on Pesticides is fundamentally intended to establish a comprehensive and robust framework aimed at effectively reducing the various risks and impacts associated with pesticide use across member states. The principles of preventive action and precaution are absolutely paramount in the ongoing efforts to halt and even reverse the alarming current trend towards increased pesticide use and dependency that has been observed over recent years. The framework established by the SUD comprises a wide range of measures designed for both general and specific purposes, which collectively address critical aspects such as training, accreditation, advisory services, and thorough inspection of equipment in relation to the general obligations imposed by the SUD. General measures are specifically intended to create a significant impact on all agricultural practices involving pesticide use within a Member State; accordingly, the adoption of these measures is not just a recommendation but rather an obligatory requirement for compliance. Specifically related to the use of plant protection products, the measures outlined in the directive include provisions for establishing pesticide-free buffer zones near protected sites, as well as encouraging the development of restructuring and economies of scale within the industry. The obligations set forth by the SUD encompass both obligatory general measures that all Member States must adhere to and optional specific measures that can be decided upon by individual Member States based on their unique circumstances and needs. This dual approach allows for flexibility while maintaining essential standards aimed at protecting health and the environment. (Hamlyn, 2015).

However, even in the absence of SUD measures, Member States' existing policies and measures may afford many of the same advantages. The SUD obligations are an agenda for action, expressing the intent to develop regulatory measures. Several of the Member States' existing measures would need to be strengthened, some developed, and others newly adopted in the implementation of SUD. Member States may argue that the SUD legislation does not encompass or make provision for measures that impinge on internal markets. The SUD obligates government authorities to use existing legal instruments to implement measures in the final synthesis of works. For example, under the SUD, regional authorities in the Netherlands could, as a contingency, adopt a state of emergency based on the existing Pesticides Act, severally confronting farmers with draconian measures and even criminal outcomes. However, such extreme measures taken against cardinally affected farmers risks a Climate-Zilt farmer retribution or a Bass stranglehold on the public administration; both scenarios are to be avoided at all costs.

11.4.5. National Policies Supporting IPM

The Food and Agriculture Organization of the United Nations (FAO) strongly advocates for the implementation of integrated pest management (IPM) as the most efficient and effective strategy for managing agricultural pests in developing countries around the world. Since the early 1990s, the concept of IPM farmer field schools (FFSs) has proliferated significantly on a global scale, generating considerable optimism for the advancement of ecological agriculture, the adoption of organic farming practices, improvements in food security, and efforts aimed at reducing rural poverty. Nonetheless, despite these advancements and the growing awareness of the benefits of IPM, the actual adoption of this approach by farmers has been regrettably slow across many nations, hindering its potential impact. There exist very few national policies designed to encourage and foster agricultural pest management methods that prioritize IPM instead of relying on hazardous chemical pesticides. Current practices commonly involve ultra-low-volume aerial spraying of monsoon fogs containing toxic pesticides intended to manage outbreaks of migrant pests, along with higher-volume applications of broad-spectrum systemic insecticides for everyday crop pest management. Consequently, in the absence of cohesive national policies that actively support the adoption of IPM and impose disincentives for the use of harmful pesticides, the widespread adoption of this more sustainable approach by farmers will likely continue to be inconsistent and limited.

National IPM policies serve as a crucial and necessary precursor to the effective development and implementation of national IPM programs or strategies and are vital for encouraging the subsequent adoption of Integrated Pest Management (IPM) practices by farmers. In order to successfully promote widespread farmer adoption of IPM methods, it is essential for countries to prioritize the formulation and execution of robust national policies that actively support the principles of IPM. These policies should be accompanied by clear goals, realistic timelines, frameworks for monitoring and evaluation, as well as dedicated budgets that align with both national investment initiatives and commitments made on international platforms pertaining to climate change and biodiversity conservation. Importantly, these policies must give particular attention to marginalized rural populations who rely heavily on agriculture not only for their food security but also for maintaining their livelihoods. This focus should be coupled with a strong emphasis on gender inclusiveness, ensuring that both men and women in farming communities can benefit equally from IPM practices. Furthermore, it is essential that these policies adopt a holistic ecosystem approach to agricultural pest management. They should actively discourage and prohibit any policies or practices that may promote or incentivize reliance on chemical agriculture. Additionally, the establishment of comprehensive pest and pesticide databases is crucial, along with the implementation of bans on high-risk pesticides that pose threats to health and the environment. Finally, effective licensing of pest control operators and pesticide applicators must be enforced to ensure that all practices are conducted safely and responsibly, safeguarding both community health and ecological integrity.

In 2002, the Food and Agriculture Organization (FAO) introduced pivotal policies and comprehensive programs that were necessary to effectively scale up Integrated Pest Management (IPM) practices, particularly focusing on the context of rice farmers across various regions in Asia. The FAO put forth national IPM policies aimed at fostering active farmer involvement in the adoption of affordable, practical, and ecologically sound agricultural practices within countries such as India, Indonesia, the Philippines, Malaysia, Thailand, and Vietnam. Among these initiatives, the FAO outlined several diverse IPM programs, which were accompanied by a dedicated program-wide implementation strategy and the establishment of a specialized IPM Facility designed to support these efforts. Across the diverse landscapes of these countries, a total of six to eight national IPM programs were initiated, reflecting a wide-ranging effort to enhance agricultural sustainability; however, it is noteworthy that none of the national policies achieved full operational status. With the sole exception of Vietnam, the minimal agricultural support budgets that are currently in place in the majority of these countries have predominantly been funneled into chemical agriculture. This unfortunate allocation has primarily supported the use of toxic pesticides, the application of chemical fertilizers, and the prevalence of monoculture cropping systems, thereby hindering the potential for sustainable agricultural development. (Pretty & Pervez Bharucha, 2015)

Chapter 12

In-Depth Case Studies on IPM (Detailed Analyses with Data & Lessons Learned)

Contents:

12.1. FAO's Farmer Field Schools in Indonesia (Rice IPM Success)

In the year 1999, the Food and Agriculture Organization of the United Nations, commonly referred to as FAO, took a significant step by launching a comprehensive Farmer Field School (FFS) program focused specifically on the production of chili peppers in the region of Aceh, Indonesia. The primary and major objectives of this FFS initiative were to enhance and improve the knowledge base and production capabilities of chili pepper farmers residing in Aceh, while also providing crucial support to the revival and recovery of the Aceh agricultural sector, which had been severely affected and devastated by conflict-related violence during the tumultuous period from 1999 to 2005.

The Madrasah Adoption/Takeover-End Contract Scheme (MATS) approach was employed as a strategic initiative. The MATS approach capitalizes on the valuable experience gained through rice Integrated Pest Management (IPM) Farmer Field Schools (FFS) that were set up over more than one agricultural season. During the initial season, the program was facilitated and organized by an independent group that was initiated by the Food and Agriculture Organization (FAO) in collaboration with various stakeholders actively involved in the farming community. In subsequent seasons, the intention was for the facilitation to be taken over by the local agriculture office of the government, referred to as the Department of Agriculture Extension and Assistance (DAEA), or potentially by a farmer organization, or even by non-governmental organizations (NGOs), or a combination of these entities, all operating under an agreed-upon contract with the DAEAs. The duration of the contract was expected to last for at least two full seasons, which would empower the farmers to gradually take over the control and management of their own fields as well as the learning processes involved. The farmers' fields were carefully adopted as FFS fields, and these were constructed on a contract basis between the facilitators and the farmers following the MATS principles. Prior to commencing the FFS, a farmer group was formed to foster collaboration, but the group initially lacked experience in effective farmer group management and in conducting learning processes ahead of the FFS sessions. The facilitators diligently designed the FFS program iteratively, continuously incorporating feedback and evolving with the farmer-relevant topics. A variety of engaging and interactive learning approaches were applied throughout the entire agricultural season, making sure to closely relate them to the farmers' daily decision-making processes and operational practices. Additionally, numerous group discussions, spirited group games, and enriching group study tours were organized to actively enhance and encourage farmer participation, ensuring that all participants were engaged and invested in the learning experience.

At the end of the season, farmers in FFS were able to directly participate in advising other farmers about what they learned, and a farmer was elected to be the head of the group. Evidence of farmer involvement was represented by farmers eagerness to arrange visits to the learning plots of other farmers. Despite the success of FFS, it was difficult to transfer the arrangement toward a full-fledge farmer-run group who will plan and execute a learning program independent of external facilitators. Farmer organizations such as farmers groups and cooperatives did not seem to be an appropriate agency to take over the FFS design and facilitation assistance, as they were also lacking of personnel, resources, and management skills (Mariyono et al., 2013).

Rice IPM Success: There is a growing and increasingly acknowledged consensus on the critical importance of pest management as a fundamental component for achieving sustainable agricultural production across the globe. Currently, crop loss attributed to pests averages an alarming 30% worldwide, with figures reaching up to a staggering 80% in certain tropical and sub-tropical regions where conditions greatly favor pest proliferation. The widespread failure of various pest-control attempts is also becoming widely recognized as a significant challenge, particularly in food-production systems that are vital for feeding the growing global population. However, recent advances in pest control methodologies have raised hopes for the development of more sustainable pest management solutions that come with fewer negative side effects and reduced environmental impact. Five methodologies that are becoming increasingly important in contemporary pest management strategies include biogeography, which studies the distribution of species and ecosystems in geographic space and through geological time; coupled human-natural systems that integrate social and ecological factors to find balance; decision and artifact systems aimed at improving decision-making processes; indicative and integrative methods that enable comprehensive assessment and integration of ecological data; and integrated pest management, which combines various management strategies for effective control. During the mid-1990s, one of the foremost causes for concern, and at that time the largest insect pest threat to rice crops, was the widespread and devastating outbreaks of the planthopper in China and several South-East Asian countries. In Thailand, despite the implementation of a variety of control measures, rice crops in certain areas remained completely devastated, leading to significant economic losses for farmers and threatening food security. It became urgent and imperative to devise effective methods for comprehensive pest management that could protect crops and ensure stable yields. As a result, Integrated Pest Management (IPM) was selected as the preferred method and strategic approach towards modern pest management practices. The aims of this paper are, first, to briefly outline the structured IPM system currently in use in Thailand, and then to present detailed, in-depth case studies that highlight some of the more successful components of the IPM system, including a particular focus on sustainability and the long-term viability of agricultural practices.

Rice is considered the most pivotal and significant crop across the globe, providing the primary food source for an astounding 50% of the world's population, making it a fundamental part of daily diets in many cultures. In Thailand, rice is not just another crop; it holds the title of being the most essential crop among the diverse 19 crops that are cultivated throughout the country. Each year, rice is planted on approximately 8 million hectares of land, making use of previously cropped and meticulously cultivated agricultural fields to maximize productivity. The agricultural sector, which plays a vital role in the economy, is significant in its contribution of 12.4% to the gross domestic product. It is particularly noteworthy that rice and rice-derived products account for a considerable 23% of the nation's agricultural exports, cementing rice's status as both a staple food and an economic beacon. However, there are concerns associated with the usage of pesticides in rice cultivation. It is recognized that the use of pesticides may inadvertently alter the dynamics of pest species, leading to an unforeseen impact on non-target species within the ecosystem. Such changes could potentially result in pest problems manifesting in different and more challenging forms. Additionally, the adverse effects of pesticides are manifold; they contribute to problems such as insecticide resistance, increased secondary pest outbreaks, the high costs associated with pesticide application, human and animal poisoning incidents, and the detrimental effects on biodiversity, which includes vital resources such as fishes, turtles, frogs, snakes, birds, beneficial insects, and soil microorganisms. Given these issues, there is an urgent necessity for a more sustainable and environmentally friendly pest-control program that can address these challenges effectively. Integrated Pest Management (IPM) has emerged as an environmentally friendly agricultural practice that emphasizes sustainability and has been successfully implemented in numerous countries worldwide. However, it is only in the last decade that IPM has gained traction and recognition in Thailand. The Thai IPM Program, which was notably adopted, drew inspiration from the Community IPM program and also integrated insights from the Doctorate in Environmental Science provided by the International Rice Research Institute starting in the year 1997. This collaboration has laid the foundation for improving pest management techniques in rice cultivation, ensuring the long-term viability and sustainability of this critical crop for both the economy and the environment.

12.2. Push-Pull Technology for Maize in East Africa

Cereal stemborers are recognized as the most significant pests affecting maize, which is the primary staple food crop in Kenya and Uganda. The impact of these pests is profound, with estimated losses in yields exceeding USD 1 billion each year across various regions, including the Americas, Africa, and Asia. Following a thorough prioritization exercise aimed at concentrating research efforts on this pest complex, there was notable involvement from the private sector in the development of sustainable and environmentally safe control technologies. As a result of these collaborative efforts, a groundbreaking approach was discovered that integrates both ecological and genetic management techniques. This innovative method, referred to as 'push–pull' technology, harnesses the synergistic potential of a repellent plant that is strategically planted both within and around an attractive trap cropping plant. The goal is to manipulate the behavior of the stemborer moths and enhance their natural enemies, ultimately fostering effective pest control measures within the maize crop. The research partner institutions played a critical role in uncovering this promising approach, which has been aptly named the 'Push-Pull Technology'. Furthermore, the efforts for technology dissemination have been efficiently coordinated through a multinational initiative aimed at spreading knowledge and practices related to this pest control strategy. Push–pull technology stands out as a prime example of integrated pest management technology, employing pest manipulation techniques to enhance an ecological process for pest control. This approach has garnered recognition, being honored as one of the World's 'Top 10 Innovations that could change the World'. The current study elaborates on how valuable research insights are effectively translated into tangible economic, social, and environmental impacts. These transformations are achieved through partnerships that build capacity at various levels, including international, national, and local spheres.

Push-pull technology has undergone significant evolution, transforming into a comprehensive and expansive platform that is dedicated to the development of the African continent. This platform is specifically aimed at ensuring not only the safe but also the sustainable intensification of smallholder maize production, which is crucial for food security. This innovative platform effectively consolidates a diverse range of participants who play key roles in its implementation. Members from the Scientific Community, various Non-Governmental Organizations (NGOs), Government Research and Extension Agencies, additional Development Agencies, the proactive Farming Community, and invaluable Private Sector Partners all come together in this initiative. Through a collaborative and participatory capacity development approach, the effective engagement of African partners within this platform has been tremendously supported and enhanced, encouraging contributions from all sectors involved. This inclusive approach has gained notable global recognition within the agricultural research and development community, emphasizing its critical importance and effectiveness in addressing agricultural issues. The practical experience that has been gathered from the application of push–pull technology is not only relevant but also widely applicable to a multitude of intricate agricultural challenges faced across Africa and beyond. This revolutionary technology holds the immense potential to provide a comprehensive global framework of solutions, which could ultimately lead to dramatic improvements in the livelihoods of one billion smallholder farmers around the world. It is a transformative initiative that addresses the specific needs of farmers while promoting sustainable practices. Moreover, platform leaders have proactively initiated meaningful and interactive discussions with various stakeholders and interested parties. These discussions aim to explore the most effective strategies for tackling the multifaceted agricultural challenges encountered in their respective countries. They are committed to understanding how push-pull technology could play a pivotal role in effectively addressing and overcoming these pressing challenges. Through such interactions, they aim to foster a greater sense of collaboration and innovation in the quest for sustainable agricultural development across the African continent.

12.3. California Almonds – Mating Disruption for Navel Orangeworm

California stands out as the most significant contributor to the global almond supply, surpassing all other regions by a wide margin. In the year 2018 alone, California yielded an impressive 2.2 billion pounds of almonds harvested from approximately 1.6 million acres of land, leading to a remarkable overall economic value of about $6.6 billion. The harvest period for almonds typically occurs from mid-August to late August when the hulls of the nuts naturally split open. The process of almond harvesting involves carefully removing the nuts from the trees, followed by a drying phase on the orchard floor. Once adequately dried, the nuts are then swept up and collected for processing. It is crucial to calibrate the shake size prior to harvesting to minimize the risk of bruising the trees and to prevent hull rot from developing. To address ant infestations effectively, orchard management practices include removing filtration slices that have been used to limit ant movement over time. This step reduces the availability of off-farm locations for ants, forcing them to seek food elsewhere. In situations where ants pose significant problems, the application of water cannons serves to drive them away from the orchards, consequently increasing the selection pressure on the foraging ants. This strategic approach contributes to reducing ant populations in the future. Integrated Pest Management (IPM) practices, such as the push-pull method and biological control strategies utilizing beneficial organisms like monarch spiders and the tiny wasp species known as Thripobius semilute, are increasingly gaining traction throughout the almond industry. The collaboration of crop advisors and regulatory bodies fosters this trend toward adopting more sustainable practices. Furthermore, the introduction of biocontrol agents, such as side-suckers, web-spinners, and girdlers, is starting to yield favorable results for almond growers in California, enabling them to cut back significantly on pesticide application. The almond industry is also grappling with ongoing chronic air quality concerns resulting from dust pollution generated by dirt roads, as well as dust from orchard removal and tillage activities. In response, numerous integrated management solutions are being rolled out, including the planting of cover crops to stabilize soil, outfitting haul trucks with hoppers and errant-capture netting, and employing temporary road watering techniques using stationary sprinkler systems to mitigate dust emissions. Moreover, the presence of pest control districts, the imposition of neonicotinoid restrictions, and various Minimum Residue Level (MRL) and product-related challenges underscore the continued necessity for the industry to explore and promote alternatives to conventional insecticides. There is an increasing push for the use of biopesticides and beneficial insect populations to reduce reliance on synthetic chemicals, ensuring a more sustainable future for almond cultivation in California. The commitment to these innovative practices not only benefits growers but also enhances environmental health and biodiversity in the region. (Payne, 2019)

Mating Disruption for Navel Orangeworm: Navel orangeworm (*Amyelois transitella*) (NOW) has increasingly emerged as the most damaging pest posed to nut crops in California, necessitating an urgent and effective implementation of alternative conditions aimed at their management. The situation surrounding pest control has reached a critical juncture, emphasizing the importance of innovative strategies. There are numerous opportunities to develop and refine new techniques specifically designed to disrupt the mating behaviors of NOW and *S. exitiosa* by utilizing various types of mating disruption technology. A wide range of dispenser types and application rates, along with diverse methods of application, have been meticulously studied. Additionally, considerations for baiting protocols and yearly maintenance cycles have been explored, leading to dispenser replacements that have also been scrutinized. All these factors have been examined thoroughly to establish the optimal conditions necessary for achieving the best performance of the mating disruption system. This system aims to effectively control the navel orangeworm, which is known for its poorly dispersive nature. The primary goals of this study are to provide comprehensive and in-depth case studies that highlight the performance results derived from these extensive studies.

The mating disruption system, which uses a specialized dispenser to release a time-staggered blend of non-saturating and saturating pheromones, is an ecologically friendly and earnestly needed method to control NOW populations in orchards with high broods. The effort started in the late 1990s, and lab studies were conducted to characterize and develop techniques and products to disrupt mating. The first semi-commercial test in an almond orchard began in 2000, and since then, the mating disruption system has been commercially available for the control of NOW in almond orchards under both grower-leased and company-operated scenarios. At present, the system is being used in over 60,000 orchard acres for nut crops, with a growth rate of 3,000 acres annually. The target pest in this section is NOW, and the focus is on the development and performance testing of the mating disruption system.

The combination of slow-releasing formulations with high-release rates has been in commercial use for the effective mating disruption of two relatively long-lived moth pest species. Under conditions that are long-lived, mating disruption has proven to be extremely successful; however, it is not particularly robust when faced with wide differences in brood length, respiration rates, and varying levels of fumigant vulnerability at an orchard scale. The introduction and commercialization of mating disruption techniques for NOW were first seen in the year 2000, and this was accomplished through the innovative use of a combination of high-release and slow-releasing sex pheromones. (Frank et al., 2020).

12.4. Biological Control of Cassava Mealybug in Africa

The Nigeria-based International Institute of Tropical Agriculture (IITA), established in the year of 1967, is a renowned and impactful non-profit agricultural research institute dedicated to improving the agricultural landscape of Africa and beyond. Utilizing innovative biological control methods, the institute undertook a significant challenge: tackling a pest that had proven to be notoriously destructive to maximum agricultural yields in cassava crops.

Consequently, efforts commenced to develop sustainable methods aimed at controlling the cassava mealybug in Africa, all without relying on harmful pesticides. Since the year of 1978, the brunt of that adaptation has frequently been bolstered by participants attending various galas and opportunities, either through the introduction of this pest's natural enemies or via training numerous agricultural institutions to replicate these successful strategies. The systematic methodology employed is clearly explained and typically unfolds in 3 detailed steps. By the year of 1988, that concerted effort marked a pivotal moment, scoring significant success with a remarkably impressive benefit: cost ratio of the cassava mealybug program calculated at an astonishing 149:1. More than 200 million individuals across Africa rely on cassava as a crucial source of food, being recognized as the chief staple food in the vast sub-Saharan African region. Notably, no extensive pesticides are required for controlling the cassava mealybug, which had been a significant problem in the past, allowing many small-scale African farmers to be liberated from the burden of this damaging pest by harnessing the power of nature itself.

The first step in biological control of a pest is to search its native home for natural enemies or consult the existing ones. The second step is to commandeer those natural enemies in the home area where they will attack a particular pest. The third step is to shoot that pest and its natural enemies intact to the new area. The first successful step was taken in 1986. The most effective parasite wasp was mass reared and sent out to various IITA stations. The Technical Package prepared and Assessment of Field Testing schemes, as the second step, were also developed. In 1987, the second step was undetected in Africa.

Cassava Mealybug, an important pest of cassava, was brought from Latin America to Africa in 1971. By 1977, it had spread extensively and caused crop losses estimated at between 10 and 80 percent. Its piercing and sucking mouthparts sucked sap from cassava plants, stunting cassava yield. Enhanced environmental biological control of cassava mealybug was patterned after the successful program against the cassava mealybug. The mechanism, predicted at 1991, is explained in two steps. An expectedly significant drop in mealybug population was observed six months after release. At Ngozi grew raw mango. Twelve mango trees severely infested by mango Mealybug were observed prior to the release of parasite wasp.

12.5. Vietnam's "Three Reductions, Three Gains" Rice Program

Vietnam, recognized as one of the world's prominent rice producers, has undergone a variety of successes as well as challenges in its efforts to combat rice pests effectively. Since the 1980s, pest management strategies have evolved significantly from a government-controlled model that heavily relied on chemical pesticides to more diversified and sustainable methods. These modern approaches incorporate farmer participatory training, enhanced regulatory measures, and the implementation of biological control techniques. This case study meticulously discusses the Vietnamese experience with Integrated Pest Management (IPM), with a special focus on the noteworthy "Three Reductions, Three Gains" rice program and its essential contribution to the nationwide up-scaling of IPM practices. The early successes of a pest management program, which was adopted at the farm level in the 1980s, are examined first, paving the way for a deeper understanding of the emerging rice pest problem that subsequently developed. This discussion includes the evolving paradigm of crop management techniques and the initial government interventions that were implemented to address these challenges. Despite the substantial progress made over the years, there remained a series of persistent challenges that needed to be addressed. This led to the adoption of a more diversified approach that sought to engage farmer organizations actively, while also extending efforts beyond the farm level to involve local governments and civil society in a collaborative manner. Through this broader engagement, the objectives of sustainable rice pest management have been pursued more holistically.

Rice is a stable food crop for the majority of Vietnamese people and the most important agricultural commodity with other important roles (Thorburn, 2015). High paddy area and yield have contributed closely to food security, the principal goal of the government since unification. Rice-quality enhancement and system diversification were introduced as new strategies to achieve economic sustainability of the rice sector. Uniting two prevalent rice production systems – the lowland wet rice system in the north-dominated by irrigated rice; and the coastal and riverine environment of the Mekong Delta disturbed by saline intrusion, flooding, and drought – a diversified ecosystem management framework was developed and piloted, showing a feasible pathway to sustain rice production in the transforming climate. By maintaining the natural and anthropogenic unproduction systems, productivity could be improved without considerable external input increase. Understanding biophysical and socio-economic structure and functions of rice agroecosystems provided insight into the ecological agriculture based strategy that the Government and the civil society organizations are implementing against climate change (Pretty & Pervez Bharucha, 2015).

12.6. Middle East's Date Palm – Red Palm Weevil Suppression Program

Date palm, scientifically known as Phoenix dactylifera, is an important member of the Arecaceae family, and it is frequently cultivated primarily for its delicious fruit. The history of its cultivation can be traced back to approximately 3000 BC, when its significance was recorded in the archaeological remnants found in ancient Mesopotamian cities, grand temples, and majestic palaces. Additionally, references to date palms appear in some of the earliest Egyptian papyrus writings. Given its extensive history, the date palm is considered one of the oldest cultivated fruit-bearing palm trees utilized not only for culinary purposes but also for various other applications. The date fruit itself holds a position as one of the most important and distinctive types of fruit cultivated in arid zones, thriving in environments where few other crops can survive. The global production of dates is estimated to reach around 7.5 million metric tons, which are harvested from an extensive area measuring approximately 154,579 thousand hectares. Cultivated varieties of date palms have been successfully grown in the arid regions of the Middle East and the Gulf Countries since 1974. However, the success of their cultivation varies considerably, which is influenced by factors such as local climatology, soil conditions, and the presence of pests and diseases that might impact growth. Date palms can be found flourishing throughout the Omani coasts, where they stretch from the northern-most region of Musandam all the way to Salalah, with a majority commonly planted throughout the interior regions. The Salalah area is particularly known for its fertility, producing a variety of palm dates that are highly valued. Thanks to the favorable climatic conditions ideal for the growth of the date palm, thousands of these palms are cultivated not only for consumer sale but also for the purpose of livestock feeding. Date plants have established a wide presence throughout the coastal areas, although there are some that can also be found inland. Among the various species, the natural Gharib and Irsad palms are often planted as boundary markers to delineate land. Not long ago, two highly sought-after varieties of Khlas date palms were imported from Abu Dhabi and Khor Fakkan in the United Arab Emirates. These newly introduced varieties have been reported to be extremely productive and are in high demand in the bustling markets of Muscat and Dubai. (M. Al-Kindi et al., 2017)(Hassan Mirza et al., 2021).

Entomological contributions are lacking, although they account for the majority of insect pest concerns in many agricultural and natural ecosystems. Date palm tree palm trees are perennial monocots in the family Areaceae on which a few pest species are observed. The Middle East's date palms host harmful pests such as conical-bore beetles, termite-weevils, whiteflies, and borers. Red palm weevils (*Rhynchophorus ferrugineus*), a pernicious pest of date palms, have emerged as a significant threat in the Middle East following the 1980s. In Oman, the economic impact of damage to date palms has increased dramatically between 2000 and 2010, with incalculable mortality rates and increased logging. *Ommatissus lybicus* Distant (Homoptera: Musidae) is a new and formidable pest of date palms in Oman that harms fronds and ultimately results in the death of the batch within a year of establishment.

***Red Palm Weevil Suppression Program*:** Throughout the expansive and diverse Middle Eastern region, the red palm weevil, which is scientifically identified as *Rhynchophorus ferrugineus* (Olivier), belongs to the extensive family of Coleoptera: Curculionidae. This insect has emerged as a notorious and highly destructive polyphagous pest that is causing widespread alarm among agricultural stakeholders. Over recent years, this pest has gained significant incursion, effectively marking it as one of the most feared and dreaded pests that severely affect both date and coconut palms, which are vital agricultural resources in this part of the world. The red palm weevil pest infiltrates various palm species, including coconut, date, fan, and oil palms, as it burrows into their soft, fleshy portions, ultimately leading to extensive damage that can be devastating for growers. Record high levels of loss in cultivated palms due to this pest are further exacerbated by the use of infected planting materials that significantly facilitate the spread of the weevil. In addition, during logging operations, the dismantling of palm trees, or even the natural death of already infected palms, a shift in infestation pressures is observed, affecting different growth stages and making this problem pervasive for both immature and mature palms. This phenomenon amplifies the extraordinary damaging capacity of this pest and poses a persistent challenge for effective management. To combat the growing threat of the red palm weevil, high doses of chemical insecticides have been introduced into agricultural practices with varying degrees of effectiveness. The literature printed in this research will undertake a thorough review and assessment of the effectiveness of various pesticides, alongside biological control methods and prevalence studies that have been conducted specifically in Saudi Arabia regarding the infestation of the palm weevil. By doing so, it aims to provide a comprehensive overview of the current situation facing farmers and agricultural professionals while also proposing potential solutions to combat this persistent threat in their cultivation efforts. (Husain et al., 2024).

With the alarming and widespread infestation of the palm weevil, an increased reliance on numerous chemical insecticides became evident in recent years, particularly insect growth regulators that exhibited a variety of modes of action tailored for different circumstances. These pesticides were employed extensively throughout palm plantations across multiple regions to manage this relentless pest that threatens agricultural productivity and biodiversity. In order to evaluate the effectiveness of these chemical treatments on adult populations, a series of rigorous pesticide efficacy studies were meticulously conducted over an extended period. These studies yielded vital and significant results concerning adult palm weevil mortality rates, both within trenches filled with treated soil and directly on the palms themselves where the pests were notably active. Furthermore, studies focusing on the residual persistence of these chemicals reported a noteworthy decline in efficacy levels for commonly utilized insecticides such as profenophos and imidacloprid, which led researchers to reconsider their usage. These substances were applied at varying rates ranging from 0.7 to 1.2 g a.i./mL in the bark and also in cut trenches, further complicating the treatment strategies employed by farmers and agricultural specialists. While broad-spectrum insecticides demonstrated effectiveness against the palm weevil, they unfortunately also precipitated negative consequences for local wildlife and had detrimental effects on nontarget microbiota, leading to serious ecological concerns that could not be overlooked. The potential impacts on beneficial insects, soil health, and surrounding flora raised alarms among environmentalists and agronomists alike, compelling scientists to scrutinize their methodologies closely. Understanding the complexity of these interconnected issues, scientists began to shift their focus toward alternative biocontrol agents, including entomopathogenic nematodes (EPNs) and fungi, as promising potential solutions to the palm weevil crisis threatening palm cultivation. Initial glasshouse trials conducted on EPNs against the pest species were first documented in Saudi Arabia and showcased extraordinarily high mortality rates among EPN-inoculated adult palm weevils, signaling a hopeful method for control without the negative repercussions seen with chemical pesticides. EPNs were effectively isolated from palmarum wire traps strategically baited with both healthy and infected palm specimens, illustrating the ability to source the natural predators of this pest directly from the environment. Following these encouraging discoveries, extensive research exploring nematode pathology in relation to the palm weevil was undertaken with great diligence. This research was thorough, carefully examining several factors such as the infectious juvenile routes of entry, the significant roles played by mutualistic bacteria in enhancing nematode efficacy against hosts, their overall persistence in the environment under different conditions, as well as the intricate mechanisms involved in penetration, making sure every possibility was explored. Moreover, through dedicated and concerted scientific efforts, promising local EPNs were meticulously selected, leveraging cutting-edge DNA barcoding techniques to ensure their effectiveness and compatibility with the target pest species. This advanced methodology was utilized to develop gnotobiotic commercial nematode strains specifically targeting R. ferrugineus, encompassing the entire journey from the initial discovery of suitable nematodes in specimens collected in the field to reaching the commercial production level aimed at significantly enhancing field releases of these beneficial organisms to combat the palm weevil infestation. This comprehensive and multifaceted research initiative aspires not only to control the population of the palm weevil but also to facilitate the restoration of balance within the affected ecosystems, highlighting the importance of integrating ecological health with pest management strategies while paving the way for sustainable agricultural practices that can endure in the years to come. (Nurashikin-Khairuddin et al., 2022)

12.7. West Africa –Bollworms in Cotton Fields

The rapid and significant growth of vegetable production across sub-Saharan Africa has undeniably led to erratic and often excessive use of pesticides by countless farmers. This alarming trend has resulted in numerous detrimental impacts that cannot be overlooked. Some of the consequences include pesticide contamination of vital resources such as water, soil, and the very food we consume. Additionally, there are long-term effects on both ecosystem and human health, which could be grave. The increasing pest resurgence and the emergence of pesticide resistance among pests further complicate the agricultural landscape. In light of these serious issues, it becomes essential to develop and promote good practices for pesticide safety as well as proactive precautionary measures that can mitigate these adverse effects. This section aims to provide critical and enlightening information about integrated pest management (IPM) through various discussions and experiences gathered from IPM practitioners' circles in West Africa. These discussions, which bring together scientists, extension agents, and farmers, facilitate a deeper understanding of both the history and the contemporary practices of IPM in vegetable production within the West African region. The actions and initiatives taken since these enriching discussions have had a substantial impact, particularly the combination of traditional pest management methods with newly acquired biological control knowledge. These integrations have been noteworthy implementations of IPM in the West African agricultural domain following the outcomes of the discussion meetings. A significant portion of this section concentrates on evaluating the impact of IPM in the West African region. In these areas, researchers and farmers have been actively encouraged to develop and adopt good practices concerning the implementation of IPM strategies. The adoption of these effective practices related to pesticides has shown immediate benefits, notably in terms of reducing pesticide residues found in vegetables. Additionally, innovative post-harvest washing methods involving sand and vinegar have been creatively adapted by women traders of fresh vegetables. These methods are utilized to effectively reduce pesticide residues during the transportation process from production sites to market locales. A vital piece of advice is to consciously avoid any long and protracted intervening actions following the introduction of IPM strategies. Participants actively involved in the implementation of IPM have demonstrated great enthusiasm as they seek to explore the next steps and develop additional activities that support ongoing improvements. However, without adequate and continued encouragement to foster the development of good practices, there is a risk that their enthusiasm may wane. Consequently, the initiatives surrounding IPM could potentially diminish, particularly among smallholder farmers who rely on these methods. It is especially crucial to emphasize that researchers, extension agents, and farmers engaged in IPM innovations require further stimulation and opportunities for investigation. Moreover, documenting new experiences will be vital for the progressive development of effective IPM practices. The aim of this paper is to encourage eager and interested young agricultural scientists and extension agents to start actively seeking and identifying good practices of pest management. This should be based on consultative discussions and intensive participatory field trials directly involving farmers and consumers working together in vegetables and fruit production. Such dynamic discussion meetings serve as powerful methods for conducting field action research that addresses socio-economic aspects of pest management. It is important to recognize that there are no clear-cut and absolute margins when it comes to pesticide regulations, food safety considerations, and the comprehensive scope of integrated pest management methodologies. (Atcha-Ahowé et al., 2015).

Bollworms represent a complex of four or more Spodoptera species that are prominently associated with cotton and a diverse range of other crops across various regions of the world. The practices associated with integrated pest management (IPM) for cotton involve thorough and season-long scouting of the entire crop specifically for signs of bollworm damage. This includes assessing square loss and monitoring bloom-feed damage, alongside the meticulous collections of eggs, worms, and natural enemies that may impact these pests. Ultimately, cotton producers who implement IPM strategies, along with their consultants, must determine specific action thresholds or a combination of thresholds that will effectively trigger necessary treatments. Adhering to rigorous scouting procedures and appropriately using treatment thresholds will lead to well-informed recommendations for corrective spray applications tailored to the needs of the crops. To ensure the successful implementation of IPM, tick and follow-up audits of compliance with critical aspects such as scouting, treatment thresholds, appropriate selection of pesticide materials, and precise application timing will be conducted on IPM cotton producers and their consultants. This structured approach is essential for maintaining high standards of pest management. In addition to the normal economic returns generated from cotton production, producers can also expect to receive bonus payments for their compliance with the IPM program. These bonuses hinge on adherence to the detailed monitoring and treatment procedures outlined, as well as the diligent documentation of their efforts and the overall scale of their operations. Moreover, a detailed case study focusing on the control of cotton bollworms over an extensive area of 20,000 acres of cotton—primarily cultivated with open-pollinated and hybrid varieties that vary in growth duration, including early, medium, and long-duration types—on a combination of sandy and loamy soils in four northern districts of the prominent cotton-growing states showcases a remarkable instance of success. This unprecedented case has proven highly effective in the practical implementation of IPM strategies to manage cotton bollworm populations efficiently.

Past experiences indicate that sowing more than 150,000 acres with a single variety and/or in a group with uniform time and method in a limited area creates problems like the outbreak of pests, diseases, and weeds. It also reduces the water availability for other crops in command areas. Therefore, in order to avoid such problems, it was planned to conduct studies with the avowed objectives of documenting the present insect pest management practices adopted by the farmers in cotton cultivation, and assessing the relative efficacy of farmer's methods and IPM practices in controlling cotton bollworms. After the experiment was permitted, care was taken to zero out conventional farmers. Using the records on insecticides, sprays, pick-up, and ginning return, cost and returns were calculated on a per-acre basis. The percentage of cost on seeds, insecticides, weeding, hoeing, and other items of expenses was similar under each management practice. The cost of picking and ginning was relatively higher under IPM, but it was compensated by an increase of 250 kg of seed cotton production. The percentage of pick-up was also high under IPM, followed by conventional, followed by farmer, and lastly crutches. The quality of the seed cotton was the best under IPM with a gin turn out of 37 percent longer length of lint fibers, better yarn count, more astudio, less trash, and better color and brightness than the other conventional methods.

12.8. Greenhouses in Europe – Vegetables Facing Whiteflies and Thrips

The utilization of active ingredients that have been registered in plant protection products or pesticides has significantly enabled Europe, over the past five decades, to successfully produce a wide variety of fruit and vegetables without facing economic losses attributable to pests and diseases. The intricate legislative and regulatory frameworks that have been established in the European Union have undergone substantial evolution, transitioning from the earlier Council Directive 91/414/EEC to the more contemporary and comprehensive Regulation EC 1107/2009. Within this framework, end-point values for the risk assessment of pesticides are meticulously defined and should be taken into account for the sustainable and responsible use of these chemicals, alongside ongoing research into alternative defense strategies. These strategies include classical and augmentative biological control methods, the use of pheromones, semiochemicals, and leveraging plant resistance as effective measures. This present paper provides a thorough review of the trajectory of fruit production in Europe, tracing back from the green revolution characterized by intensive fruit production systems that garnered attention until the 1990s and the corresponding results of these practices. It also addresses the emergence of secondary problems that began to confront the period of prosperity, recent challenges in the field of agriculture, and pest management strategies that are being adopted toward achieving more sustainable fruit production systems. This shift is notably occurring in the context of simpler orchard environments, primarily achieved through the cultivation of better-trained human resources who are equipped to meet the demands of modern agricultural practices. (Damos et al., 2015).

High levels of productivity were granted by intensively managed systems under temperate climate and the wide use of pesticides and other chemical inputs. A decrease of pesticides has been achieved in apple with respect to the 1990s, nevertheless economic losses due to pest and disease attack increased as well environmental hedges near or in orchards. Gradual substitution of pesticides and once monthly or less application frequency allowed a smooth switch towards modern solid cultural practices that avoid previous problems of apple growing. Mild warning signals of new invasive pests and re-emergence of old ones approach with apprehension the satisfaction of consumers in Europe on food safety and healthiness.

The need for sustainable pest management strategy is widely recognized, however still poorly implemented in fruit growing at whole orchard level since it has to be applied in less complicated environments, both culturally and biologically. Knowledge of each particular orchard and rapid adjustment of the management details on the detected presence of pests and beneficials is necessary for a successful pest–beneficial interaction, but it requires more time and trained human resources, presently inadequate in European conditions. Even a decade of well-funded research in components and systems have not provided the quick solutions of simpler orchard environments. Progress toward sustainable pest management should be based on knowledge of ecology and biology of the pathogen, pest, beneficial, crop and environment interactions and higher understanding of fruit production complexity.

Vegetable producers in Florida have encountered substantial challenges that have made it increasingly difficult to manage the pervasive problems posed by western flower thrips alongside outbreaks of the tomato spotted wilt virus. The alarming and persistent spread of these invasive pests and harmful viruses within the realm of fruiting vegetables continues to pose significant threats to agricultural productivity and economic stability for farmers. This troubling situation has endured despite the modern implementation of integrated pest management programs, which have been designed to be more effective. These programs have been fortified with considerable investments that have been made in advanced greenhouse production systems. This investment aims to bolster the structural and operational aspects of vegetable production. Moreover, experiential knowledge that has been gained from efforts to establish essential components of integrated pest management in the realm of ornamentals might prove quite beneficial in developing and positioning an integrated approach specifically tailored for effectively handling the myriad issues posed by these persistent challenges in fruiting vegetables. In addition, insecticide resistance is a critical issue, as it complicates the problem further by rendering traditional control methods less effective. The adverse effects on non-target organisms that are inherent in many of these control approaches, as well as the overall effectiveness of pesticide applications in mitigating these stubborn pests and viruses, provide a complex backdrop of factors that need to be carefully considered when managing and putting into action integrated pest management programs. This multifaceted situation underscores the urgent need for ongoing research, innovative strategies, and collaboration among stakeholders to foster sustainable agricultural practices that ensure the viability of fruiting vegetable production amidst these relentless challenges.

Ornamentals and vegetables require different approaches to minimize damage from dangerous pests. Group management plans that minimize costly crop loss would be desirable and feasible. Considerable expenditures, long-term investments, and a concerted team management effort are necessary to develop successful systems. This case study presents insight gained while conducting research in Florida fruiting vegetables and ornamentals to identify integrated pest management components that might be useful in developing and implementing a comprehensive approach to managing these pests and viruses in melons and other fruiting vegetables. Specific strategies for conserving natural enemies, scouting and deciding when and how to implement control measures, and research needs arising from case studies are discussed. The strategies presented are based on commonly perceived botanical properties used or planted to repel these pests, recommendations to adjust cultural practices or plant biology to deal with invasives and viruses, and scouting methods implemented in ornamentals adapted to fruiting vegetables.

Chapter 13
Challenges of Integrated Pest Management

Contents:

13.0. Introduction

Pest management through the reduced use of pesticides has captured significant attention and has been a focal point of discourse and research over the past three decades. The need for a reduced dependence on pesticides, combined with the recognition of the dangers posed by improper use, is viewed as an absolute necessity to remedy the adverse effects on our environment, biodiversity, human health, and biological control agents that play essential roles in our ecosystems. It is now widely acknowledged within the scientific community and among stakeholders that a systematic and integrated approach to pest management, with pest economics placed at its core, would effectively address the fundamental question surrounding the management of pests. Integrated pest management is, in fact, not a singular solution; rather, it is a systematic methodology that relies heavily on the generation, analysis, and monitoring of information. Its successful implementation must be contemplated at all operational levels, ranging from individual production units to broader agro-ecosystems. Moreover, the empowerment of farmers, policymakers, and all relevant stakeholders is a prerequisite for its effective implementation. All participants within the gateway communities of pest management—including farmers, extension workers, researchers, scientists and educationists, as well as policy-makers and NGOs—need to be actively engaged and empowered in the multifaceted process of implementing integrated pest management strategies. Such collaborative efforts can lead to more sustainable agricultural practices that benefit both people and the planet. (Parsa et al., 2014)

A substantial number of different pesticides are commonly utilized in numerous countries around the globe. However, considerably less focus has historically been directed towards thoroughly evaluating the impact that these pesticides have on the environment as well as on human health. The enduring ecological damage inflicted by pesticides, particularly those chemicals that have been applied indiscriminately since the conclusion of World War II, is now increasingly being recognized and cannot be overlooked any longer. Numerous scientists, researchers, academics, and non-governmental organizations (NGOs) are energetically calling for a comprehensive review of pesticide usage practices. Meanwhile, policy-makers are consistently highlighting the pressing need to diminish dependence on these potentially harmful chemicals. Various scientific studies are increasingly exposing how pesticide-polluted ecosystems exhibit behaviors that are strikingly different from those found in non-polluted ecosystems, subsequently causing significant problems for both human health and the environment. A growing number of countries, whether they are developed nations or those still in the process of development, are in the process of formulating and implementing new laws that seek to limit and manage the use of pesticides more effectively. Nonetheless, a crucial aspect that remains largely overlooked is how to systematically minimize dependence on pesticides and address the detrimental effects that have already manifested in ecosystems and communities. This represents the most formidable challenge currently confronting integrated pest management (IPM) practices.

13.1. Lack of Awareness and Education

One significant and pressing challenge that hinders the widespread adoption of integrated pest management (IPM) approaches is the prevalent lack of awareness and comprehensive education among farmers. Many farmers are generally not aware of the various aspects and benefits of IPM, leading to a correspondingly low intention to incorporate these practices into their pest management strategies. For farmers to make effective changes to their pest management practices, they require thorough education to understand the array of options available to them for managing pests, as well as to appreciate the potential long-term advantages associated with IPM systems. It is imperative that concerted efforts are made across all regions to educate farmers thoroughly about IPM, presenting it as a viable and sustainable alternative to the pest management tools they currently employ. The clear implication here is that improving education and increasing awareness can play a vital role in enhancing the likelihood that a greater number of farmers will support and engage in initiatives dedicated to promoting IPM practices. By providing education focused on the objectives of IPM, we can help reduce significant educational barriers that currently deter farmers from adopting IPM. Regularly demonstrating the projected benefits of integrating IPM practices, such as the potential for substantial pesticide savings and reduced net economic costs associated with other agricultural inputs, may also prove beneficial in encouraging adoption. Even offering farmers minimal training or increasing their awareness regarding IPM can yield substantial effects, as it facilitates informed and conscious decision-making regarding pest control methodologies. This challenge is particularly pronounced in developing countries, where farmers often have limited access to training alternatives that promote "sustainable agriculture pest management." Additionally, the linked issues of lack of education and awareness greatly influence other aspects of education and technology adoption, such as the perceived ease of using specific technologies, their compatibility with existing practices, and the observability of their effects. Beyond individual consultations, it would be prudent to organize group training sessions aimed at equipping farmers with knowledge about potentially effective pest management approaches at their disposal. Ultimately, farmers are the decision-makers when it comes to evaluating whether they believe the proposed projects will successfully reduce costs and minimize the use of highly toxic pesticides in their pest management practices. Notably, the current landscape reveals that IPM is among the least understood options available to them, underscoring the necessity for outreach initiatives aimed at educating farmers comprehensively about various integrated management strategies. Generally speaking, extensive education regarding the wide range of pest management approaches is absolutely vital for their successful adoption, ensuring a more sustainable agricultural future. (Parsa et al., 2014).

13.2. Initial Costs and Investment

While Integrated Pest Management (IPM) may lower the costs associated with pest control, it's essential to recognize that some of the most effective interventions require a high initial financial outlay. A fundamental principle of IPM is to start with lower-cost methods and gradually progress towards more advanced biological solutions as the populations of natural enemies and antagonists naturally increase over time. This approach contrasts sharply with the use of synthetic pesticides, which often kill pests indiscriminately and can foster resistance in pest populations over time, making control more challenging. In contrast, biocontrol methods and biopesticide options typically require a certain period to become effective due to the nature of density-dependent regulation, which relies on the presence of sufficient natural predators. Farmers need some level of assurance that their investments in IPM will not jeopardize their harvests or result in reduced income in the short term. Unfortunately, vicious cycles are frequently observed: farmers may hesitate to adopt these methods due to perceived risks, and as a result, the failure to adequately control pest populations can lead to even greater losses in production. Beyond the broader investment costs associated with implementing IPM strategies, integrated practices typically encompass several elements that necessitate material inputs. For instance, field-diversifying practices require the procurement of additional seeds, while interspersed traps must be both purchased and maintained diligently. Habitat-intensifying practices may necessitate the introduction of new plant species, while degree-day prediction often relies on complex software models that must be operated effectively. Moreover, competing-pest inoculation requires an adequate pest reservoir to ensure the success of the intervention. These investments must be supported and justified by the anticipated value of the final crop and a solid understanding of available market opportunities. For newly introduced crops or product types, the financial returns are often uncertain, creating obstacles that make the necessary initial investment challenging for farmers. It remains unclear whether IPM can effectively shift mindsets toward making existing higher-cost pest-management practices more capital-intensive in nature, as change is often met with resistance. Consumer awareness surrounding the risks associated with pesticides tends to be uneven across different regions, and the lack of appropriate infrastructure can result in local markets favoring high-risk pest-management strategies that are less sustainable. Experts have reported facing significant difficulties in the enforcement of contracts with local-level buyers, further complicating the implementation of IPM techniques. Even when targeted biological or cultural regulations are introduced, increased international competition can continuously pressure growers to revert to pesticide-intensive regimes to remain competitive. Consequently, some farmers may opt to remain outside the IPM framework altogether, missing out on potential benefits. Notably, IPM has been credited with explaining the decline in pesticide consumption in China, the challenges of high pollution levels in local environments of India, the self-restraining movements observed in farming practices abroad, and the reasons why farmers may refuse to consider trading off fungicide use or opting for transgenic resistance gene options under extreme conditions. (Pretty & Pervez Bharucha, 2015).

13.3. Complex Pest Monitoring and Data Analysis

Complex pest monitoring and various data analysis techniques are increasingly required in Integrated Pest Management (IPM) practices. In parallel, there is a growing demand for advanced data analysis techniques that can efficiently mine data for a deeper understanding of processes. This understanding is vital for making informed management decisions in multi-pest systems, where multiple pest species may coexist and interact. Consequently, models designed for simulating pest generations must also address external stresses such as drought and nutrient deficiencies that can impact pest dynamics and agricultural productivity. These sophisticated simulation models necessitate high-quality input data, alongside accurate forecasts regarding land use, weather patterns, and management practices. However, sourcing these essential input data has proven to be particularly challenging, especially in developing regions where data availability may be scarce or unreliable. This issue often creates a bottleneck that hampers effective pest management strategies. As a direct result, the entire user group that relies on these management models experiences diverse treatment outcomes due to considerable differences in the overall research capacity of various regions. For instance, the full integration of Geographic Information Systems (GIS) and remote sensing applications, which are critical for effective complex pest monitoring, remains limited in many developing areas. This limitation can be attributed to both resource constraints and implementation challenges. Within these trade-offs lies significant potential; bridging the disparate expectations and resource inequalities during the design of data adaptation and analytical tools could prove beneficial for enhancing IPM practices on a broader scale. The difficulties associated with current pest monitoring and forecasting approaches when it comes to implementing solutions in complex ecosystems have been increasingly highlighted within the research community. Present pest monitoring efforts are largely hampered because they rely solely on a linear and homogeneous spatial extent of the pest populations. Such a restricted approach is necessary as the existing modelling techniques, despite their attempts to incorporate environmental considerations, primarily focus on individual pests. They, therefore, tend to ignore the myriad of other biotic interactions that may take place in these ecosystems. Co-evolution among species has been a fundamental process shaping pest communities, allowing them to adapt to similar environmental conditions and thrive within overlapping time scales. Nevertheless, there remains a clear gap in understanding both the effectiveness of current models against these relatively simple pest communities and the potential advancements in pest monitoring approaches needed to address significantly more complex systems. Furthermore, the prohibitive data input requirements necessary for model scaling, along with the reduced time and spatial frames associated with pest monitoring efforts, have created a discussion concerning the conditions found in maturing countries. As the demand for data continues to grow for both modelling and monitoring approaches, there is a recent surge of interest in data augmentation techniques that aim to alleviate some of these challenges. Most importantly, it is essential to consider the significant roles of time and space in effective scaling of pest management strategies. While learning time frames, whether they be measured in hours, days, or months, are commonly understood and thus adaptable in practices related to both pest modelling and monitoring, a common time frame for essential spatial considerations remains elusive and is yet to be successfully established. (Batz et al., 2023; Pope & DeWitt, 2005).

13.4. Limited Availability of Biological and Organic Controls

Biopesticides should absolutely be regarded as a vital and integral component of a comprehensive integrated pest management (IPM) framework or, in other words, as an essential element of a biopesticide-focused crop protection strategy. This multifaceted approach greatly promotes long-term sustainability in agriculture by significantly decreasing reliance on traditional chemical pesticides, which can often lead to detrimental effects on the environment and surrounding ecosystems. Biological control, which encompasses the strategic use of living organisms or natural mechanisms to manage pest populations, has long been recognized as a crucial element of a holistic IPM approach. In fact, biological pest management can be considered one of the earliest sustainable pest control methodologies adopted, long before the widespread advent of synthetic chemical pesticides that dominate agricultural practices today. It is paramount to acknowledge that IPM is not a static system; it is a continuously evolving discipline that actively adapts to emerging challenges and innovations as they arise in the agricultural landscape. Given this dynamic characteristic, any sustainable crop protection strategy that thoughtfully incorporates biopesticides ought to be embraced as a form of integrated pest management. The development of new biopesticide products is currently occurring at an unprecedented pace, reflecting a rising interest in more sustainable agricultural practices that are less harmful to the environment. Nonetheless, the adoption of these biopesticides has been notably sluggish. This slow uptake can be attributed to a variety of hurdles and challenges, such as the ongoing difficulty of broadening the overarching concept of IPM itself, a widely spread lack of awareness about the effectiveness and efficiency of biopesticides compared to their chemical counterparts, and the persistent competition these biopesticides face from conventional chemical alternatives that are often more accessible, readily available, and less expensive. Furthermore, issues relating to compatibility within existing agricultural systems and the fragmentation of market opportunities can further deter farmers from making the vital switch to biopesticide-based solutions. To date, comparatively little consideration has been devoted to thoroughly understanding the validity and viability of integrating biopesticides into current agricultural practices, particularly when the ultimate goal is to establish their significant role in forming sustainable biopesticide-based crop protection programs. For instance, when a farmer contemplates transitioning from using a single type of chemical pesticide for managing the problem of pests to adopting a more comprehensive IPM approach, they must meticulously evaluate and consider which combination of pest management tactics to strategically employ. Selecting the most effective portfolio of strategies is an exceedingly complex process that requires in-depth analysis and careful planning. Farmers, by their very nature, typically seek to minimize the number of different control tactics they apply while simultaneously striving for the greatest possible pest management efficacy. This scenario raises pivotal and thoughtful questions: Should a new biopesticide tactic be added to their existing control portfolio, or would it be significantly more advantageous to replace a current tactic with this new one? In multiple cases, it is indeed feasible and practical to substitute a conventional synthetic chemical pesticide with a biopesticide without causing disruption to the overall structure and efficacy of the existing IPM system. However, it's crucial to note that IPM tactics may be synergistic and interconnected, which signifies that implementations of one tactic may bolster or enhance the performance of others within the integrated system. These various interconnected dynamics contribute to the challenges faced in effectively replacing traditional synthetic pesticides, thereby complicating the shift toward adopting an IPM strategy that readily integrates biopesticides. Additionally, the concept of pesticide 'lock-in' further reinforces the notion that the adoption of newly developed pest control technologies tends to favor those tactics that closely mimic the characteristics and behavioral patterns of incumbent pesticide technologies. It is certainly worth noting that some of the most successful biopesticide products currently available on the market, such as the microbial formulation of Bacillus thuringiensis (Bt), bear a significant resemblance to conventional chemical pesticides in terms of their application methods, user-friendliness, and effectiveness against pest populations. (Chandler et al., 2011; Dunn, 2017).

13.5. Resistance to Change

There are numerous reasons, or oftentimes excuses, for not implementing Integrated Pest Management (IPM) strategies, even though it is widely accepted in agricultural circles that there exists sufficient evidence to demonstrate that IPM effectively reduces losses incurred from pests while concurrently enhancing financial returns for farmers and growers. Unfortunately, IPM is not frequently advocated by universities or agricultural extension workers, primarily due to the extensive training that individuals must undergo in order to fully understand and properly apply IPM methods. Furthermore, the significant time investment required for this multifaceted and comprehensive control approach can deter many from its adoption. Many farmers and growers tend to favor immediate and traditional pest management products that offer straightforward solutions; they often seek out season-long pest management tactics that promise ease of use. Additionally, cash or grain dealers may face detrimental consequences if they actively promote IPM methods, which can lead to hesitance in endorsing these more sustainable practices. Consequently, as a result of these challenges, there exists a relatively limited market for the development and implementation of IPM solutions. When there is some market presence for IPM, it primarily relies on restricted, scheduled pesticide sprays, which can result in unwanted pest resurgence alongside a myriad of environmental issues that threaten the ecological balance. (Parsa et al., 2014).

Resistance to pesticides by pests and resistance management programs can delay or greatly reduce the effectiveness of this exit strategy for conventional control methods. Resistance development is normally quite slow, especially if an effective resistance management program is adopted (M. Matten et al., 2004). However, in many first-world crops, pest resistance development has led to the loss of pesticide options such as the organophosphates, diazinon, and methyl bromide. The need for alternative approaches to insect control is urgent because second-generation biologics are slow to develop or impractical due to limitations in movement and dispersal. Multi-pronged pest management tactics are preferable, because insecticide applications to control a pest typically provide some level of control of the entire pest community, which may resurrect detrimental secondary pests.

Pest movement either from untreated to treated areas or up the population growth and food webs are other key considerations. In one-crop areas, even if pest control with insecticides were successful, ecological barriers limit crop movement as crops are only grown in spring or summer. In multi-crop areas, pest movement across fields and cropping systems is evident and provide avenues for pest interventions, but usually not until crop damage occurs. Additionally, plant physiology and horticultural methods of pest control can be empowering, encourage further action, and present complementary and alternative interventions. Pesticides have historically been so safe to humans and the environment. They have acted as a crutch to farmers, and relying on their availability will be problematic.

13.6. Environmental and Climate Factors

Climate change will disturb the predator-prey and parasitoid-host interactions and alter the balance between insect pests, natural enemies, and host plants because of alterations in their synchrony. The impacts of climate change on pest management using biopesticides and challenges in the management of biopesticides in a changing climate will be discussed. Climate change is predicted to lead to an increase in the abundance and distribution of many species of insects that cause economic damage to crops. Climate change is also predicted to increase the inflow of pests into new areas. Numerous complex interactions and interrelated changes caused by climate change will occur that will impact plant protection and food security, and crop protection measures must be adaptive to these changes (Sharma, 2013). There will be an increase in variability in insect damage as a result of climate change. Higher temperatures will make dry seasons drier and may increase the intensity of rainfall, making wet seasons wetter than at present. Widespread adverse consequences on crop protection and food security are anticipated because of climate change. Changes in climatic factors, such as temperature, relative humidity, CO2, pH, rainfall, and soil properties, have also shown varying effects on the efficacy of synthetic pesticides. An increase in temperatures and ultraviolet radiation and a decrease in relative humidity may render many of these control tactics less effective. Numerous factors contribute to this increased variability of damage, including higher overall pest populations, fluctuating conditions that favor pest species, more unpredictable seasonal effects, and crop exposure to multiple pests. High pest numbers, exposures to multiple pest species, and crop damage could shift local economic thresholds toward a more aggressive stance on pest management. Higher temperatures are predicted to facilitate higher population growth rates, enhancing pest species abundance, cropping systems allowing for more successful pest recruitment, and increased food resources. On the other hand, altered environmental conditions could also result in pest declines, within populations experiencing substantial heat stress or faced with food resources that become less favorable. Changes in seasonal averages of moisture could increase the frequency of years with drastically dry conditions but could also contribute to irregular flooding. This could reduce crop resilience, which could be exacerbated by pests directly attacking root systems or infesting fruit and foliage. In addition to changes in seasonal averages, as climates become less predictable, pest populations could be exposed to labile temperature regimes with more rapid fluctuations and peaks that could exceed a pest species' tolerances. Food webs are sensitive to increases in environmental variability and more irregular conditions, thus the overall response to climate change in the face of pest management would be much less predictable (H. Ziska et al., 2018).

13.7. Regulatory and Policy Constraints

The successful adoption of IPM is often influenced by external regulatory and policy constraints or incentives directed at specific aspects of IPM programs. Regulations or policies preventing the availability (or increasing the cost) of certain pest control options can limit the ability to choose effective IPM approaches. Conversely, regulations and policies promoting the adoption of one or more complementary pest control tactics can increase the likelihood that IPM will be successful (R. Matten et al., 2008). Well-conceived regulations can provide the incentives to increase the successful adoption of those elements of IPM. These incentives can be short term or long term.

Increasing stakeholder vigilance regarding potential problems in the use of high-risk pest control options is increasingly likely. These concerns can lead to the restriction of or outright ban on the use of these high-risk technologies. The precedent set with the ban on the use of certain substances for soil fumigation continues to influence the way in which pest control technologies are developed and managed in the global marketplace. Regulatory bans on these pest control options are most likely if they are used by growers in developed countries. Data and information regarding potential negative consequences of pest control options used by growers in developing countries are often limited or absent, and hence are much less likely to elicit a regulatory ban or restrictions.

The imposition of such regulatory processes influences the ability of growers to assess the pest control options available and to select the pest control approach best fitting their needs given the local context. As pest control options are lost due to high regulatory hurdles, both the economic viability and the environmental sustainability of crop production can be threatened. In such a scenario, the goal of providing food security for a growing world population can be compromised. As pest control options are lost due to the imposition of stricter regulations and consequently higher production costs, the continuous challenge of the pest becomes more costly. The situation becomes one in which fewer pest control tools are available, but more pest control is needed. Thus, for those growers that can access the remaining tools and be economically viable, the pressure to use them is great.

13.8. Scalability and Coordination Challenges

The integration of pest management on a large scale can be significantly more effective when it is coordinated across a much greater area than is typically possible on a single farm. For instance, the suppression of pest populations through methods such as the mass trapping of adult beetles, the inundative release of natural enemies, or the coordinated release of pheromone-based mating disruption might be successfully implemented across an entire landscape. However, without a well-organized and unified effort among all stakeholders, the overall effectiveness of these innovative approaches may be considerably reduced. For example, the mating disruption strategy for controlling codling moth populations in apple orchards was initially quite successful and demonstrated excellent results. Nevertheless, as time went on, the population of codling moth began to recover, which was likely due to the failure of even a few growers to properly renew their pheromone dispensers at the appropriate time. This situation highlights the critical need for consistent and timely action among all participants in a pest management program. As the scales of pest management increase, the organization of cooperative action becomes increasingly challenging and complex, requiring strong communication and collaboration among farmers and pest management practitioners. (Grasswitz, 2019).

At landscape and regional scales, the opportunities and concerns regarding research and extension collaborations differ somewhat from those at smaller scales. While it might be feasible to develop treatment protocols for the use of biocontrol agents or habitat manipulation to enhance agronomic systems and to create an expert cadre, the number of discipline-specific experts available to conduct the field evaluations and test the efficacy of these measures is limited. Likewise, the funding committed to such efforts can be a limiting factor. Coordinated decision-making, made in a timely fashion, among groups of growers that span greater distances is likely to be problematic. Nonetheless, there would be opportunities, as well as economies of time and funding, to conduct regional pest-predator surveys and modeling evaluations of pest-habitat manipulations at a temporal scale that would not be feasible regionally.

The impact of large-scale producers on integrated pest management implementation differs significantly based on the specific situation. In some cases where management is coordinated across a landscape, the direct competition with large-scale producers is less pressing than the need for addressing the pest problems posed by such competitors. For example, the burgeon population of feral honey bees in large-scale fruit-production regions may prevent the ability of small-scale orchardists to utilize easy weeds like dandelion for crop pollination. In this and similar cases where small-scale producers find themselves under threat from larger competition, strategies need to be developed to prevent small producers from being entirely driven out of production. Alternatively, it might be beneficial to improve the ability of larger-scale operations to integrate pest management with heavy machinery and tight production schedules.

13.9. Long-Term Commitment

The absence of specifically allocated funding aimed at the effective implementation of education and training initiatives, combined with the thorough planning of Integrated Pest Management (IPM) strategies at the community level, has resulted in many valid opportunities diminishing rapidly into obscurity, mere moments after the initial establishment of work plans. The development of these comprehensive IPM plans is, without a doubt, a fundamental necessity for the successful and widespread application of effective, environmentally sustainable pest management techniques. Yet, as time goes on, the initial thrill and steadfast commitment to deploying a new and innovative approach to pest management have gradually declined, culminating in a scenario where previous proactive measures stall amidst relative inactivity. Numerous agencies, researchers, and organizations have voiced concerns regarding the long-term dedication required to carry out these meticulously devised plans, and sadly, the essential funding or provisions necessary for this ongoing effort are alarmingly scarce. This predicament represents a continuous challenge that continues to beset many locales. Comprehensive pest management, whether it be centered on IPM, Integrated Crop Management Practices (ICMP), or an even more aggressive methodology, is inherently a costly and labor-intensive endeavor. It demands a unified and sustained effort to accomplish effectively, with funding being essential to this undertaking. As a result, the plans are thoroughly detailed, educational and training initiatives are arranged, and yet, in the absence of significant allocated resources, the entire process starts to noticeably slow down. This deceleration tends to linger longer in areas managed by non-federal entities, yet the glaring absence of targeted funding will inevitably cause a faster decline in the pace of development than was originally foreseen. As new urgent priorities emerge within local jurisdictions—priorities that redirect critical focus away from pest management—the ongoing initiatives to educate, staff, and train personnel concerning the plans that have been previously formulated inevitably slip further into the shadows. At the same time, the requirement for tailored solutions to urgently address more pressing pest management challenges becomes increasingly apparent. It is widely recognized among numerous federal, state, and local agencies that, on a broader scale, the propensity to adopt a comprehensive approach to pest management tends to be significantly greater during the early stages of any particular initiative. However, the sustained commitment needed to uphold educational measures and to cultivate and manage the essential relationships that allow the IPM strategy to progress, adapt, and remain pertinent will always present a substantial challenge. Historical insights illustrate that the situation is typically easier to navigate when operational complexities are minimized, and when ample resources are available to back the ongoing efforts underway in various communities. (Nowierski & Meyer, 2008)

Chapter 14

The Future of IPM

Contents:

14.0. Introduction

In the mid-1970s, the concept of Integrated Pest Management (IPM) began to gain wider currency in California. Research into the Red Imported Fire Ant (RIFA, Solenopsis invicta) invasion and turf disease problems undermined reliance on chemical pesticides as the primary control method. The problem definition was expanded. Since the 1950s, successful control efforts had targeted turf insect pests and turf diseases. IPM research then focused not only on expanding prospects for biological control, but also changing the socio-economic and political conditions affecting urban turf ecosystems in the state's diversified economy. The researchers recognized that crop pest problems were intertwined with non-crop pest problems in California's landscape, and that turf and ornamental crops were not just agricultural crops deserving of control efforts.

Some efforts, despite official state policy, were not pest management actions at all. Outreach and mechanism development subsequently targeted urban and semi-urban pest management. The alternative concept of "Integrated Urban Pest Management" was proposed in the early 1980s. Empirical studies into socio-economics, pest complaints, practices, and civic institutions followed. Changing university missions highlighted basic IPM research deficiencies.

Beyond landscape urban ecosystems, funding mechanisms for action attempts surged. As California came to dominate the world's economy, all states tried to establish inter-agency, multi-disciplinary IPM approaches for the state's unique ecosystems. This story may resonate with those seeking to create public-private partnerships for Integrated Pest Management (IPM) and other innovative mechanisms in exotic invasive pest and disease pest management implementation initiatives for the next millennium. Be cautioned and forewarned, nevertheless, as the California story ended differently than others. In some cases, the systems built retracted like an overused rubber band, leaving a pattern of disparate aims and hopes akin to an urban landscape filled with patches and foci of development, but devoid of corridors. In California, IPM came to be regarded as a non-issue in pest management decision-making and practice. Causative factors include paradigmatic shifts in California's economy, and a fortuitous state drought that further inflated expectations for "E-Government" to turn it around. Integration and outreach had come to a gradual end. Federal and state funding mechanisms targeted site-specific disciplines more exclusively than systems; public-private partnerships became the province of the privileged few; and viable species fled the beleaguered urban landscape. The response that followed is offered here as caution for IPM advocates elsewhere and as a rallying point to coalesce the strength of science and practice across disciplines to counter the unfolding crisis involving socio-economics, the environment, and the food and fiber supply (M. Nowierski & J. Meyer, 2008).

14.1. Climate Change & Emerging Pest Threats

Climate change, in conjunction with the escalating levels of carbon dioxide (CO_2) present in our atmosphere, is set to have an incredibly profound and profoundly far-reaching impact on a wide variety of agricultural practices, horticultural endeavors, and forestry initiatives. These significant alterations in our environment will not only affect agricultural practices and cultivation methods but will also have significant repercussions on the intricate and delicate biotic interactions that transpire among the rich and diverse array of flora and fauna that inhabit our planet. As these critical environmental shifts unfold progressively, it becomes increasingly essential that Integrated Pest Management (IPM) strategies undergo extensive revision and updating in order to effectively accommodate and tackle these transformative changes and challenges we are bound to face. In parallel with these notable and necessary improvements in environmental conditions, the hierarchical rankings of various plant pests are poised to change dramatically, reshaping our understanding and management strategies. This significant shift will consequently bring forth the emergence and proliferation of new insect pests, a variety of invasive weed species, and a range of disease-causing organisms that present potential threats to crops, as well as the overall health and productivity of plants. Therefore, there is an urgent and pressing necessity to acquire a deeper, more nuanced understanding of how varying climate variables, coupled with increased and heightened CO_2 concentrations, exert influence over the complex interactions between plants and pests. This vital knowledge base will serve as the foundation for creating innovative and robust risk assessment models that can accurately predict the forthcoming challenges and pressures that the agricultural sector will encounter over the next quarter of a century. The multifaceted interactions between changing climatic variables—such as fluctuations in temperature, alterations in precipitation patterns, and increased levels of CO2—coupled with the performance and behavior of plant pests, will constitute the central focus of the upcoming presentation. By delving deeply into and comprehending these intricate dynamics, we can better prepare ourselves to navigate the complexities of effective pest management in the face of a rapidly evolving environmental landscape, ultimately ensuring sustainable practices that will benefit agriculture and contribute to food security in the long run, fostering resilience in our agricultural systems.

Low-cost IPM strategies have been devised for the management of a relatively new pest of cotton, Palmer amaranth (Amaranthus palmeri). It is categorized as a noxious weed in 19 southern and Midwestern states. Palmer amaranth is a glyphosate resistant (GR) weed that rapidly spread in the Southeastern United States following the introduction of Roundup Ready cotton and is now present or rapidly spreading throughout other states with cotton and soybean crops. Palmer amaranth can produce in excess of 1 million seeds from a single plant in one growing season. It has many adaptations allowing rapid evolution to environmental change, including the use of multiple outcrossing reproductive systems and rapid natural and artificial selection. Exotic biotypes also exist from different geographic and ecological regions which have different genetic backgrounds and correspondingly different responses to management practices. Palmer amaranth is rapidly evolving resistance to Group 14 and Group 15 herbicides in Georgia, Alabama, Minnesota, South Dakota, and Nebraska. This rapid adaptation has resulted in the need to develop and implement herbicide and non-herbicide management options immediately and urgently. However, due to hundreds of years of inaction in the face of consistent chemical selection pressure, Palmer amaranth is likely to continue evolving resistance to many effective post-emergent, pre-emergent, and non-selective herbicide classes and is also likely to evolve resistance to additional physiological classes of herbicides on the market. (H. Ziska et al., 2018)

14.2. Sustainability Metrics and Long-Term Impact Assessment

A significant and pervasive issue that is currently plaguing modern agriculture involves the detrimental impact of pests, pathogens, diseases, and weeds, which collectively damage approximately 30% of the world's crops each and every year. These losses translate into an astonishing estimated cost of about $540 billion annually on a global scale. Among the vast array of more than 87,000 pesticide products that are available for sale, many of them carry concerning side effects. These side effects encompass various health risks, including neurotoxicity, reproductive toxicity, mutagenicity, immunotoxicity, and carcinogenicity. Moreover, the use of pesticides is contributing to declines in biodiversity, negatively affecting pollinators, and harming soil health. These adverse effects extend beyond individual farms, damaging essential ecosystem services that are crucial for the functioning of our environment and economy, with real global economies suffering losses valued at a minimum of $1.5 trillion each year. In light of these alarming impacts, there is an urgent and critical need to explore and implement methods that will reduce our ongoing dependence on synthetic pesticides. This can be achieved through sustainable farming practices, the management of alternative agricultural inputs, and the adoption of holistic agroecosystem approaches that not only aim to enhance agricultural productivity but also prioritize the health of our ecosystems and the safety of our food systems.

Integrated Pest Management (IPM) relies on understanding the ecology of plants and their pests. It is the deployment of a variety of methods of pest control designed to complement, reduce, or replace the application of synthetic pesticides, combined with simultaneously managing diseases and weeds. IPM is also a form of sustainable intensification—a process of managing farming inputs and practices that over the long term both improves productivity but at the same time improves wider ecological, economic, and social environments. IPM incorporates the simultaneous management and integration of tactics including non-chemical, chemical biotechnological, engineering, and socio-political interventions aimed at controlling plant pests. One of the most striking features of IPM is that it is notoriously resilient over time. The evolution of pest populations in response to management, coupled with the fact that pesticides are inevitably diluted, degraded, and leached away, leads to the conclusion that the IPM challenge is never done.

IPM can be defined at different and complementary scales. At the broadest, IPM is a science encompassing the study of all types of organisms that affect plants. At finer resolutions, an IPM can also refer to the methods that control a specific population of pests and pathogens. A common feature of IPM that is at heart a biophysical and management-oriented practice, is a systems understanding of plants and the selection and deployment of interventions that exploit that knowledge to augment the unwanted with desired effects, feedbacks, and dynamics. Historically, IPM did not proceed through a particular theory and has many origins in academic, farmer, governmental, private, and global perspectives of candidate interventions, their scientific basis, and systems understanding. Research and extension networks emerged from and collaborated with these organizations on information exchange and capacity building for knowledge and theory production and pest management practice adoption or improvement.

14.3. Integrating Traditional Knowledge with Modern Practices

Integrated Pest Management (IPM) is specifically designed to develop a more comprehensive understanding of sustainable crop protection and management systems that are both effective and environmentally friendly. This approach enables practitioners to take into account a wide array of factors, including cultural practices, various methods of crop rotation, effective soil management techniques, the development and use of improved crop varieties, genetic characteristics that can enhance yield and resistance, compatible species that could be utilized within crop systems, natural enemies of pests and biological control methods, as well as the application of machinery and metal work for enhanced agricultural efficiency. Furthermore, professionals within this field focus on monitoring options and establishing thresholds for intervention, as well as anaesthetic methods for pest control and the utilization of chemical control strategies that incorporate new generation selective products designed to minimize environmental impact. Practitioners within the realm of IPM typically employ an inherent heuristic approach to case thinking and problem-solving strategies while coordinating their efforts with the surrounding natural environment, the socio-technical landscape, and relevant policy environments that govern agricultural practices. They recognize and appreciate that all prescribed IPM-based methods tend to carry more appreciable downside risks when compared to methods that are either IPM-neutral or actively IPM-blocking. It's important to note that stakeholders involved in this field can differ significantly in their perceptions regarding the credibility of vital determinants linked to IPM as well as the various categories of methods employed. Such in-depth and detailed information is essential for crop-protection scientists, as it helps them to effectively guide their strategic planning and implement necessary interventions, including potential policy changes. Moreover, this knowledge can be crucial in mitigating highly visible, imminent, and often insurmountable challenges related to foreign parasitic pest restraints, especially in an ever-evolving, complex environment that requires adherence to established suppression protocols and practices.

Indigenous biocontrol agents and traditional knowledge should be integrated with research-derived IPM methods and knowledge built from border biosecurity experiences. Crop protection is viewed as integral to other component systems of agriculture and should build broadly symbiotic relationships with them. Knowledge is seen as synergetic and evolution-based and seeks to develop a fuller understanding of crop protection as it pertains to specific indigenous biocontrol agents, crops and traditional agroecological environments. Stakeholder groups include custodians/scientists of indigenous biocontrol agents, raising indigenous-non-crop biota custodians, knowledge coding facilitators, crop group networks, and surveillance/technical-scaffold trainers to protect knowledge infrastructure and code high-value foundational knowledge for educational material to counter-the extremes of the biocontrol agent knowledge gulf. The delivery agencies of provision resources include the cooperatives of custodians of indigenous knowledge facilitators of composition disarmed insurgents of knowledge-providers and custodians of genetic resources as a means to wider. Delivery agencies are encouraged and plans of e-government structure and content with expert real-time translation of customary intellectual property rights pro forma.

14.4. Technological Innovations in Pest Management

Over the last three decades, there has been an impressive array of significant technological advances that hold great promise for enhancing pest management practices and significantly reducing the financial and environmental costs associated with pest control methods. The improvements in the speed and affordability of computers and sensors have paved the way for the development of far more powerful monitoring and forecasting systems, which have a transformative impact on how we approach pest management. Additionally, the remarkable progress made in computational power has facilitated the improvement of existing chemical formulations, the development of new technologies that enhance control efficacy, and the determination of their effects on non-target organisms that might be affected by these interventions. Advancements in molecular detection methods have become increasingly reliable, sensitive, and widely available for the detection and identification of various pests and pathogens. These cutting-edge technologies encompass a wide spectrum, ranging from individual devices that can be operated by growers in the field to sophisticated Automated Weather Station networks capable of covering large agricultural areas efficiently. However, leveraging these innovative tools requires not just access to technology but also a solid understanding of both the technology itself and pest behavior, as well as integrated pest management programs, in order to truly realize their potential benefits and ensure effective pest control outcomes. (A. ANDERSON et al., 2020)

On-farm monitoring or engineering systems instrumented with small weather stations provide forewarning of pest risks and can potentially notify users of events such as hot spots that should be scouted. Monitoring, scouting, and testing are basic components of many pest control strategies so such technologies can enhance efforts in this area. Improved geographical information systems allow growers to integrate different datasets for a comprehensive understanding of pest, crop, and environment interactions. Also, once a pest outbreak has been confirmed, better communication tools are available to alert others to the problem.

Control efficacy technologies include all potential pesticides and other means of pest control; their evaluation is thus a huge and growing task. Efforts to automate this process using robots are promising for niche opportunities but beyond the capability of current technology for most uses on large-scale crops. As manufacturers develop new chemistries, each can be modelled to determine the likely resistance risk, biopersistence in the environment, and non-target impacts with a range of crop-centred phenotyping; high throughput screening and imaging equipment are available for this task. Each evaluation could cost many million dollars and take at least a decade.

14.4.1. Smart Pest Monitoring Systems

Technological advancements are increasingly being applied to effectively develop innovative smart pest monitoring systems designed specifically to control and manage pest bugs through the use of insect-monitoring automatic trap devices. Integrated Pest Management (IPM) represents a crucial and holistic approach to the ongoing challenge of pest control. Pest Monitoring Systems (PMSs) provide an invaluable capability to track and monitor the status of pest populations in real-time. This real-time monitoring capability enables the timely and appropriate deployment of control countermeasures, which are essential for maintaining ecological balance. A well-designed PMS must consist of various components, including traps specifically engineered to capture insects, sophisticated sensing devices capable of detecting the captured insects, a robust transport layer to transmit detection results efficiently, a pest monitoring service designed to process detection results and provide user-friendly interfaces, and a reliable storage service dedicated to storing long-term detection results in an organized manner. Moreover, smart pest monitoring systems are currently under active development, aimed at enhancing their capabilities with features such as improved insect-monitoring algorithms, advancements in trap designs, and expedited system deployment while achieving lower operational costs through the integration of cloud computing and crowdsourcing technologies. Innovative remote trap devices have been introduced, focusing primarily on the detailed design of the trap case as well as the development of smart trap detection algorithms. The design of the trap case is ingeniously crafted to facilitate the fall of captured insects into a designated storage box via a trap door, which simplifies the process of collecting the captured insects significantly. Various smart trap detection algorithms utilize an array of output images depicting the initial state of the trap case, the rotation angle of the trap door, temperature records, and the specific triggered frame for detection purposes. These algorithms rely on advanced machine learning models to continuously improve the effectiveness and efficiency of Pest Monitoring Systems (PMSs). (López et al., 2012).

IPM, or Integrated Pest Management, is an important and effective pest control approach that plays a critical role in promoting sustainable agricultural development practices. With the implementation of IPM, various pest bugs are managed and controlled through numerous environmental and biological means. This approach significantly reduces the heavy reliance on chemical pesticides, fostering a healthier ecosystem and minimizing potential harm to beneficial organisms. The pest population status requires continuous monitoring to effectively implement diverse IPM strategies, and often, a multitude of pest monitoring traps are utilized specifically for pest control purposes. For many common agricultural pests, the tools employed for monitoring typically include yellow rectangular sticky traps, meticulously designed such that the insects entering the trap cannot find a way to exit. Consequently, the number of pests successfully caught within these devices is directly proportional to the overall pest population density in the area. Over time, it has become standard practice in agriculture to periodically collect crucial monitoring data from these traps, allowing farmers and agricultural practitioners to adjust their IPM control countermeasures accordingly based on the data gathered. However, due to the inherently passive nature of pest traps, the manual collection of pest data is still required to determine the exact number of captured pests. This necessity highlights a growing need for smart and innovative pest monitoring solutions that would enable pesticide-free monitoring while minimizing or altogether eliminating the requirement for human intervention in the process.

14.4.1.1. Sensors and AI-Driven Predictions

In recent years, the ongoing plague of pests has significantly impacted the agricultural economy, as these pests consume more than one-fifth of the total harvest in the entire production cycle. The struggle to combat this relentless plague has led to the adoption of innovative technologies designed to enhance the effectiveness of pest detection and prevention measures. Among the various technological advancements, sensor and computer vision technology stand out as two of the most crucial and transformative solutions being utilized to reduce the extensive losses caused by pests in the realm of online agriculture. These sensors have the capability to foster a smart farming environment by leveraging numerous material parameters such as temperature, humidity, light intensity, soil moisture, soil temperature, and a multitude of other factors. This intelligent farming method is exceptionally proficient in detecting the losses incurred from pests and gathering comprehensive data regarding pest history through the establishment of monitoring systems equipped with multiple sensors that strategize data collection effectively. Simultaneously, the computer vision component plays a vital role by accurately identifying the species of pests through advanced image classification and other sophisticated image processing techniques. Through the application of deep learning technology, a computer can achieve an almost human-like detection rate for identifying pests. This remarkable capability enables the pests to be marked and meticulously recorded within databases, facilitating the collection of crucial information about pest growth, consumption behaviors, and survival patterns. As a result, this gathered data empowers agricultural specialists to target and eliminate these pests far more efficiently through the judicious use of pesticides and other control measures, thereby safeguarding the integrity of the agricultural systems and increasing overall productivity. (Blanco-Carmona et al., 2023)(Batz et al., 2023)

When building a smart farming detection system, it is crucial to carefully consider the specific types of sensors to utilize and also the most effective locations to deploy them for optimal performance. For effective temperature and humidity detection, using readily available off-the-shelf modules can be quite efficient and practical. However, it is essential that these sensors are appropriately shielded from the direct impacts of rain, intense sunlight, and accumulated dust, which can affect their accuracy. One key consideration to keep in mind is that soil moisture sensors generally do not perform well in sandy soil environments. Research has shown that most capacitive soil moisture sensors tend to fail in this type of soil, resulting in inaccurate readings and unreliable data. Therefore, it is recommended that soil moisture sensors be specifically installed in clay soil, where they tend to function optimally and yield more consistent results. Furthermore, other parts of this sophisticated system must maintain a level of transparency; alternative forms of soil moisture sensors or any other relevant type of sensor could potentially be developed to gather valuable information that feeds into the computer vision component of the system. By incorporating a sufficient number and a wide variety of sensors, comprehensive data regarding losses due to pests in varying soil types and their respective growth cycles can be systematically collected. This data will ultimately facilitate the creation of effective pesticide action plans by allowing for accurate forecasting of both the timing and the precise amount of pesticides that should be applied, maximizing crop protection while minimizing environmental impact.

14.4.1.2. Decision Support Tools for Farmers

Pests are living organisms that actively compete with various plants for the finite array of available resources. There exists a continuous interaction between the environment and pests, which enables them to repeatedly adapt to various physical and biological conditions found within that environment. This ongoing adaptation process has resulted in significant yield losses in major food crops, ultimately impacting the entire food security system across different regions. With the recent growth and adoption of high-yielding varieties and monoculture cropping systems, the scenario of pest management has undergone a profound transformation due to the introduction and release of exotic pests into new habitats. Agricultural practices such as mono-cropping, extensive irrigation, shortened fallow periods, and increased reliance on chemical inputs have collectively created favorable conditions for the rapid build-up of pest populations. Complementing these agricultural practices, the advent of modern, novel insecticides has transformed the current frameworks of chemical control measures, consequently altering the behavioral patterns of insect pests and, in many cases, rendering traditional chemical pesticides increasingly ineffective. Plant pest infestations can occur almost instantaneously and tend to progress progressively over time. Because of this rapid progression, the deployment of a strategic preventive protocol in the face of pest threats is crucial; without it, crops are likely to suffer significant losses. Once a crop loss occurs, the course of action shifts towards reactive management strategies that respond to the damage that has already been inflicted. To combat this challenge effectively, the timely dissemination of information regarding pest build-up is critical. This not only makes essential pest-related information readily available at the farmers' doorstep but can also act as a significant deterrent to both domestic and alien pest invasions. Such detailed information empowers farmers to make informed, timely decisions regarding potential preventive measures. It is essential to regularly assess pest populations in diverse agro-ecosystems due to the inherently fuzzy nature of pest information, which often includes varying minimum, maximum, and average pest values. Moreover, visualizing the pest build-up scenario in both time and spatial dimensions is important for the development of Barricade Protocol considerations, which subsequently facilitate effective planning of pest management actions. However, for crops that are cultivated widely, accurately assessing the pest build-up scenario proves to be a challenging endeavor through traditional tabular pest data derived from low-tech sensors. These conventional pest information systems often lack the necessary sophistication to provide timely and actionable data. In contrast, advanced pest information systems are equipped with a digitized pest management information base that includes a variety of plant protection measures and information kiosks designed to assist farmers. In an API x-of-y survey, the assessment of pest build-up relies heavily on expert opinions and is conducted only after establishing a satisfactory linkage between the estimated crop sown areas and comprehensive management packages tailored for those crops. Predictive parameters such as width, free-edge, and a 10 km buffer zone are meticulously calibrated to facilitate daily updates regarding pest assessment. The daily surveillance of pests, disseminated through GIS mode at various ICT Kiosks, is widely shared with farmers through community radio channels, ensuring that desired pest control measures are directly tailored to the current pest state, weather influences, and disease development patterns. Decision support systems specifically designed for assessing pest build-up scenarios and integrated pest management ecosystem applications function to disseminate real-time pest assessment and management recommendations effectively. These systems have been innovatively developed for early detection programs that establish links between farmers and their respective state agriculture or horticulture departments, thereby aiding in prevention and recovery activities related to pest management. The dissemination of daily progressive pest build-up information in real time through community channels holds significant relevance in enhancing export quality control standards for crops such as pumpkin, okra, and other fruits and vegetables, which require stringent low pest incidences to ensure superior quality in buy-sell markets. By implementing customized pest management information, farmers can work towards achieving better outcomes in their agricultural endeavors, thus strengthening overall food security systems. (Kumar Singh, 2018)(Penot et al., 2018).

14.4.2. Drone Technology and Precision Agriculture

In recent years, unmanned aerial vehicles, commonly referred to as UAVs or drones, have been increasingly deployed across various industries, particularly in the field of agriculture. These innovative tools are now utilized for the collection of imagery which is crucial for effective crop and pest monitoring. The commercialization of UAV technology for aerial imaging has progressed rapidly, making these devices widely available to farmers and agricultural professionals. Moreover, drones that are specifically equipped for spraying purposes, alongside their imaging capabilities, are also readily available on a global scale, further enhancing their application in agricultural settings. The versatility of UAVs does not stop there; they can be seamlessly integrated with ground-based vehicles such as ground-rovers and advanced all-terrain robotic vehicles. This integration is especially beneficial for pest monitoring in horticultural fields, where crops are often planted in rows and large ground-based vehicles can struggle to navigate. By employing UAVs in such scenarios, farmers can achieve a level of precision and efficiency that was previously unattainable. Advanced image processing techniques applied to images captured by UAVs facilitate a robust method for weed mapping. This approach not only allows for accurate identification of weed infestations but also holds great potential for future integration with UAV sprayers. Consequently, this synergy results in site-specific weed management, enabling farmers to target their interventions more effectively and optimize their use of resources. As the technology continues to evolve, the role of UAVs in agriculture is set to expand even further, delivering enhanced productivity and sustainability to the industry. (E Hunter et al., 2019)

As the spatial aggregation of weeds increases, the efficiency of UAV management practices has correspondingly risen, marking a significant advancement in agricultural methodologies. Weed mapping utilizing UAV imagery is emerging as a novel alternative resource for the research community, the agrochemical industry, and farmers' organizations, enabling the development of site-specific weed management strategies tailored to particular fields or conditions. With the surging global demand for food and fiber, alongside steep increases in agricultural input costs, it is imperative for the sector to optimize productivity while minimizing resource expenditure. To achieve this, agriculture must innovate and strive to accomplish more with fewer inputs and resources. Precision agriculture stands out as one of several promising approaches being thoroughly explored and promoted to enhance efficiency in farming practices. Among the many tools available within precision agriculture, unmanned aerial vehicles (UAVs), often referred to as drones, have recently attracted significant attention from various stakeholders in the agricultural sector. These UAVs, which are being offered at relatively modest prices, are now being marketed to growers not only for the purpose of scouting fields and collecting data but also for the aerial application of critical inputs such as pesticides, herbicides, and essential nutrients. Furthermore, UAVs that are equipped with advanced imaging sensors provide farmers with specific and intricate levels of detail that facilitate informed decision-making regarding vital tasks like irrigation, fertilization, and pest management interventions. UAVs that are capable of spraying can effectively crop dust substantial areas for several thousand dollars; however, numerous questions remain regarding their overall efficiency and effectiveness in various agricultural applications. (C. Nunes, 2023)

14.4.3. Advances in Organic Pesticides

The growing threat that insect pests pose to agricultural crop yield, overall quality, and efficient storage solutions has led to an urgent need for the advancement and implementation of sustainable, organic pest control measures. These measures strategically utilize the natural resources and biological diversity available to us. In order to effectively compile an inventory of potential compounds and develop them into practical products, it is essential to have a thorough understanding of several key factors. These include the mode of action of a given compound, its potency against targeted pest organisms, its compatibility and safety with non-target organisms, as well as its environmental impact and overall fate once released into the ecosystem. A progressively greater number of innovative methods aimed at the prevention and management of pests and diseases are emerging and are currently being rigorously examined and evaluated across Europe. Novel or new active substances for pesticides are being actively developed, increasingly utilizing plant extracts and entomopathogenic fungi. Furthermore, biopesticides are being designed for introduction into spray tank mixtures alongside existing chemical products, allowing for a more integrated approach. Other noteworthy innovations are taking form in the development or trial phases of several devices aimed at enhancing pest management. For instance, the Trap and Kill system is specifically designed for the targeted control of adult insect pests, while the Predictive Modeller serves as a cutting-edge tool for disease forecasting and decision support. Additional advancements include slow-release pheromone products tailored for door and window foliage, as well as the innovative use of plant roller blinds designed for effective light exclusion. These advancements present significant opportunities for refined pest management practices within established frameworks such as integrated pest management (IPM) and organic agriculture production systems. The impact of these innovations will be felt not only by the farmers who implement them, but also by the chemical supply companies operating in the crop production sector, particularly within the value chains of tomatoes and peppers. It is crucial for farmers to effectively communicate both the positive and negative real-world impacts that these innovations have on their practices to future innovators and researchers in the field. Additionally, it is important to address possible impediments that may hinder the introduction or broader acceptance of the innovations currently under development. Such challenges must be considered during the framework development phase, concurrently with further development of the innovations themselves. Among these impediments are considerations of cost-effectiveness, practicality of products and practices, overall knowledge and awareness, regulatory burdens that can impede progress, resistance to change, and potential effects on crop quality that must be consistently monitored. The notable advantages of utilizing Case Studies in this context are that they are fundamentally underpinned by extensive research projects, which provide a comprehensive written and audio-visual record of meaningful discussions with farmers and various stakeholders. This approach not only enriches understanding but also enhances collaborative efforts in the pursuit of effective pest management solutions. (R. George et al., 2015)(Umetsu & Shirai, 2020).

14.4.3.1. Plant-Based Solutions

Natural pesticides derived from various plant species and utilized for managing pest populations are commonly referred to as biopesticides. These biopesticides are essentially bioactive compounds that plants synthesize as secondary metabolites during their intricate metabolic processes. In current agricultural practices, biopesticides offer a promising alternative to the widespread use of synthetic pesticides for effective pest control and disease management. The extensive reliance on synthetic pesticides has been associated with numerous detrimental side effects, which has spurred interest and innovation in developing safer alternatives sourced from nature. One of the notable advantages of biopesticides is that they typically exhibit minimal adverse impacts on non-target organisms when compared to traditional chemical pesticides, making them a safer option for the environment and biodiversity. Furthermore, the presence of multiple bioactive compounds with known or unknown mechanisms of action within plant extracts significantly limits the chances for the development of pest resistance. This characteristic is critical in ongoing pest management strategies, as it allows for more sustainable practices. Plant-derived biological control agents have proven to be effective, demonstrating high efficacy against various pest species while simultaneously exhibiting a low probability of causing resistance development among target insect pests. In the realm of pest management, products that originate from plants are classified into distinct categories such as biopesticides, repellents, and additional plant-derived substances that can disrupt mating, oviposition, or feeding behaviors of pests. Specifically, plant essential oils are recognized for their significant biological activities, which include insecticidal efficacy against pests infesting stored products, their capacity to act as repellents, and their ability to deter oviposition. This is particularly relevant for controlling indigenous, exotic, and invasive insect species that hold considerable economic importance across various agricultural sectors. Moreover, the potential for mixtures of plant essential oils, each displaying varying levels of efficacy, has been recognized as an innovative solution for managing insect pests in niche crops that are economically valuable. The importance of plant-derived biological control agents as viable alternative non-chemical control methods has been thoroughly summarized, emphasizing their effectiveness against a diverse array of pests and diseases. Additionally, it is essential to consider the compatibility of these biological control agents with their natural enemies, ensuring that their application does not disrupt existing ecological balances. The discussion further extends to the potential of utilizing essential oils, whether individually or in mixtures, for managing significant agricultural pests as well as the pests associated with the storage of fruits and vegetables that often lead to food waste. Understanding their compatible application in various contexts is crucial for maximizing their effectiveness in pest management strategies. (R. George et al., 2015)

Plant-derived insect pest management products (PDPPMPs) could help to avoid resistance development in target pest species, address growing public concern regarding the safety of chemical pesticides, minimize adverse effects on nontarget organisms and the environment, ameliorate pest management in niches, and promote human health and wellness by protecting food quality (Kumari et al., 2022). Determining the efficacy, mode of action, and likely compatibility with natural enemies of PDPPMPs is imperative before their introduction. Three groups and multiple categories of PDPPMPs have been delineated: (i) biopesticides that kill target pests by directly acting on them, (ii) repellents that reduce pest damage, and (iii) PDPPMPs that inhibit mating, oviposition, or feeding of insects. Currently available artisanal PDPPMPs categorically as biopesticides, repellents, and other bioactivity have been described and innovatively discussed. Biopesticides are currently available from seed oils, essential oils, non-volatile plant extracts, neem seed extract, living fungi, and bacteria. Useful modes of action include volatile and non-volatile components of PDPPMP based on various plant parts. Validation and verification of utility in commercial cultivars are needed to broaden the scope of PDPPMP products through the commitment of academia, industry, and government. Plant-derived products other than essential oils would help to maintain the efficacy of essential oils over longer periods.

14.4.3.2. Biodegradable Chemical Alternatives

- ***Overview of Chemical Alternatives***

Sustainable replacements for hazardous chemicals are increasingly in demand across a variety of sectors. Though many potential substitutes exist, few have been technically or commercially developed. Candidate chemicals must compete against established substances and be demonstrably sustainable or bio-derived. Suggestions for new substances must thus consider the properties and applications of competitive chemicals, as well as sustainability metrics. Biobased chemicals with evidence of being more sustainable throughout their life cycle are of particular interest. A common tenet of most "greenness" metrics is that renewability is often considered a requisite condition for sustainability. This notion appears to be a motivator in many recent studies on the viability of alternative bioproducts, highlighting a relative lack of focus on bio-based substitutes that are simultaneously greener. Work in this area is timely given the unprecedented attention currently afforded to bio-based substances, and has the potential to advance the state of knowledge in both green chemistry and chemical substitution fields.

Prior studies aim to identify evidence as to whether readily available biobased replacements exist. Emphasis is placed on already commercialized compounds with open literature, suppliers, and applications similar to existing chemicals. A comprehensive collection of potential bio-sourcing strategies for commodity chemicals is presented; however, the analysis does not explore the potential sustainability benefits of substitutes. This work aims to identify biobased replacements found in the most comprehensive commercial and scientific organizations databases. Particular emphasis is placed on competition with hazardous substances. Existing evidence for replacements is presented and subsequently consolidated to highlight gaps in knowledge that warrant further investigation.

It is also noted that "sustainable" is a broad term that can be interpreted in a variety of ways as applied to authenticity or chemicals. Given the current absence of legislation or regulatory definitions governing the commercial structuring of materials or chemicals as sustainable, focus is placed on the renewability of bio-derived substitutes as a requisite condition for sustainability (Wu & T. Maravelias, 2019).

- ***Definition of Chemical Alternatives***

Since ancient times, humans have both consciously and unconsciously altered their local environments as a foundation for the settlement of human populations and the development of communities and societies (Wu & T. Maravelias, 2019). There is evidence that early farming communities in the Americas and the Eastern Mediterranean conducted organic chemicals and pesticides to enhance soil quality, actively cultivating plants for medicinal purposes, cleaning and hygiene, and recipes for foods and beverages. One of these communities, in the valley adjacent to the Great Lakes, employed innovative strategies using multiple plants to produce smoke, steam, and fog that were transported to capture minerals in the atmosphere and prevent degradation of their territory. Stoic preventions failed worldwide recently as chemicals spilt into the environment, and chemicals manufactured for applications in diverse sectors have become persistent and problematic. Chemicals entering the environment through anthropogenic activities become pollutants when they bioaccumulate in biologic cycles that affect organisms. Chemical alternatives, characterized as non-toxic, biodegradable, and low bio-accumulation, are alternatives to existing chemicals acknowledged as hazardous.

Efforts to mitigate the effects of hazardous chemicals have focused on regulatory change, research into current alternatives, and innovations to render current chemicals safer. Research into fast to evolved biodegradable organic chemical alternatives produced of the same elements has been disregarded. Organic compounds that are presently non-toxic, biodegradable, or low bio-accumulation do exist, yet their prevalence, utilization, and potential specialized and proprietary applications in replacement of hazardous chemicals are arguable. Chemical alternatives will be fast to evolved biodegradable organic compounds employing breakdown pathways and target organisms long absent from the environment and bio-accumulated compounds. Target candidate chemicals in pre-industrial groups are vindicated, alongside analysis of their fast chemical degradation and leaching potential in comparison to the persistent target chemicals employed industrially. The motivation is to introduce and inspire novel ideas for an innovative future in chemical design and selection that is applicable in diverse sectors.

- ***Importance of Sustainability***

The 21st century is an era of sustainability in which products and processes need to meet the triple bottom line of economy, society, and environment (Maliki et al., 2022). Emerging global restrictions aim to achieve a sustainable society. These regulations seek to advance this sustainable revolution via an innovative economy based on bioprocesses, bioresidues, and bio-resources. In view of these challenges, chemically synthesized products, ranging from air/pesticides, medicines, and plastics, need to be substituted by biodegradable and biocompatible alternatives. The ambition of this review is to motivate academic and industrial research on innovative non-petrochemical chemical alternatives, products, and other applications.

The chemicals, polymers, and products obtained from biomasses and waste/biomass residues have received widespread attention regarding their sustainable and biodegradable properties. They protect the environment and public health from harmful and non-degradable petrochemical derivatives, preventing the formation of micro and nano-plastics. Such chemicals are biodegradable and bio-compatible compared with chemical alternatives. Highlighted sustainable chemicals and products, from which developed and innovative commercial products can be derived, have the potential to impact modern technologies and consumer markets profoundly (Summerton et al., 2019). Non-toxic, biodegradable surfactants to replace the petrochemical-derived surfactants employed in household hygiene, personal care, food, textiles, and oil industries have been developed. With the rapid growth of the perfumes market, propagation-free floral fragrances, wood fragrances, citral, and essential oils have received widespread attention as green replacements for current petrochemical perfumes. Edge-stable, optically clear bioplastics are designed from biopolymers such as gum, starch, and spider silk to replace polystyrene windows in electronic devices. These chemicals are advantageous to health safety and star technology industries.

- ***Types of Biodegradable Chemicals***

There are two general kinds of biodegradable chemicals: biobased and fossil-based. On the contrary to natural polymers, which are exposed to degradation in the environment but are hard to recover for reuse, such as plastics and rubbers, biodegradable chemicals are chemical substances that do not remain in the environment forever, are used worldwide, and can be recovered for reuse in virgin substances after decomposition. Synthetic biodegradable chemicals have been developed from petroleum or coal sources. Fossil-derived biodegradable chemicals rapidly decompose into natural polymeric materials, which do not remain in the environment. Therefore, they do not add to the global carbon footprint caused by the deliberate introduction of carbon from buried fossil resources. These chemicals can basically be produced from both fossil resources and biomass. Biobased biodegradable chemicals slow down the increase of global warming by reducing carbon dioxide emissions in the producing-fossil-sources process (Nakajima et al., 2017). They have high potential as sustainable and environmentally friendly alternatives to fossil-derived chemicals.

14.4. Policy & Research Needs for Global IPM Adoption

Effective pest management is of paramount importance to the global systems that support food production, and the various pest management practices in use are closely intertwined with the kinds and amounts of agricultural inputs employed. These inputs range widely and include crucial resources such as fertilizers, soil amendments, irrigation techniques, and genetic inputs, all of which are vital to ensuring adequate crop yields. In addition, these practices are closely linked to broader issues pertinent to the growth and distribution of food supplies across different regions. In today's world, sustainable food production has emerged as an urgent global challenge, particularly as the expansion of land designated for food crops approaches its limits. This creates a renewed focus on the significant inefficiencies that characterize our current food production systems, prompting discussions about viable and essential alternatives to improve these systems. Such alternatives are often categorized under the broad term "agricultural intensification." This term encompasses the simultaneous adoption of a suite of agronomic practices designed to maximize outputs from the same area of farmland. Additionally, there exists a previously untapped potential that could lead to substantial increases in food production solely through the application of Integrated Pest Management (IPM) techniques. There is a pressing need for heightened research efforts and policy backing that support increased investments in and the widespread implementation of IPM methodologies on a global scale, particularly in developing countries that often face the greatest food security challenges. The IPM coalition is well-positioned to leverage notable events, including periodic high-profile conferences, that emphasize the vital role of IPM from varied perspectives—environmental sustainability, human health, and economic viability. Collaborative initiatives in developing nations, often in partnership with philanthropic organizations, have begun exploring alternative solutions such as bioherbicides derived from medicinal plants, the use of switchgrass and toasted seeds of bush pigweed as effective bioherbicides, and the cultivation of pest-resistant varieties such as Bt and T-LGA cotton. Moreover, there is ongoing research focused on providing practical recommendations for the non-chemical control of insect pests, including those affecting the aquatic species Aponogeton madagascariense, commonly known as Madagascar water lily. Despite the documented successes presented in this paper and the continuous efforts of numerous stakeholders to promote biocontrol methods for spontaneous and effective utilization on and around farmland, the overall rate of establishment success for these initiatives remains disappointingly low. This underscores the importance of further investment and strategic support for enhancing the efficacy and reach of such sustainable practices. (Pretty & Pervez Bharucha, 2015)(Parsa et al., 2014).

14.5. Collaborative Research Initiatives

Integrated pest management (IPM) represents a multifaceted technical and regulatory challenge that must be addressed urgently. To effectively navigate this landscape, it is essential to gather fundamental and thorough information regarding pest biology, interactions between pests and plants, and the various technologies available for pest management. This crucial information must be meticulously synthesized, carefully integrated, and thoughtfully packaged into practical and effective approaches that will significantly reduce the risks associated with pesticide usage. The challenge of technology transfer in this context is to guarantee that key stakeholders across multiple domains—specifically agriculture, human health, and the management of natural resources—are fully aware of the latest advances in IPM technology management as well as the diverse regulatory avenues that can empower their decision-making processes. Furthermore, agro-ecosystems must be designed to accommodate and address the inherent shortcomings associated with most existing food production technologies. Therefore, it is necessary to develop high-yielding pest-resistant germplasm that not only retains competitive vigor but also demonstrates responsiveness to critical management inputs. In the realm of regulatory frameworks, practical criteria are essential for evaluating and accounting for the potential adverse ecological effects that food production systems can impose during registration decisions for new technologies. The establishment of heterospecific assessments pertaining to the efficacy of pest management technologies is paramount for the key players involved in the IPM technology transfer efforts. In terms of geographic focus, initial targets for these endeavors should include the developing systems found within the Caribbean region and the southern hemisphere nations. Several collaborative research initiatives are currently operational and dedicated to addressing these challenges. For example, the focus of the Invasive Terrestrial Animals Project (ITAP) encompasses a range of issues, including invasive terrestrial vertebrates, invasive pests that affect human habitations, and invasive pests along with diseases impacting crops, domestic animals, wildlife, and trees specifically within terrestrial ecosystems, while intentionally excluding weeds and aquatic organisms from consideration. The primary purpose of this document is to raise awareness about the ongoing crisis in systematics within federal agencies and what these implications mean for biosecurity measures within the United States. Within this context, there is an account provided of the fate of the Great Lakes Seaway snail observed in the central United States. Alarmingly, the USA is experiencing a rapid decline in both terrestrial and freshwater biodiversity at a deeply concerning rate. For decades, the unimpeded introduction of approximately 20-30 non-native species on a monthly basis has led to a sustained net increase in the number of non-indigenous species inhabiting these ecosystems. Subsequently, native and culturally significant species have become increasingly vulnerable to extinction, primarily due to aggressive competition and a series of interactions mediated by bio-invader pathways that facilitate such ecological disturbances. (M. Nowierski & J. Meyer, 2008).

14.6. Where Will IPM Be in 2050?

The International Organization for Biological Control (IOBC) was founded in France in the year 1955 with a significant purpose: to promote the adoption and effective use of biological control in the management practices for pests on a global scale. Over the years, the IOBC evolved and expanded into a comprehensive network of various international commissions, each dedicated to specific aspects of biological control and pest management. One notable commission formed was the Commission on Integrated Control, which was specifically commissioned with the goal of promoting Integrated Pest Management (IPM) methodologies and practices worldwide. At the request of the Food and Agriculture Organization (FAO), the Commission undertook the task of preparing a detailed set of open-ended papers, which eventually coalesced into a significant and informative collection entitled "Integrated Pest Management." This influential collection was officially released in the year 1980 and comprised an extensive overview of IPM on a global scale. It included individual papers that addressed various crucial topics, such as the economic aspects, ecological considerations regarding pest management, and detailed studies on insect pests, weeds, nematodes, and diseases that affect agriculture and forestry across the globe. The insightful papers within this collection were authored by a total of 38 contributors, many of whom gained international recognition and acclaim in the years that followed, although some were not yet well-known at the time they wrote their papers. Among these, a particularly significant paper titled 'An operational definition of integrated pest management' emerged as a defining moment for the global IPM movement. This landmark paper laid the groundwork for future discussions and developments in the field. Building upon the foundation provided by this critical paper, the FAO requested a United Nations Intergovernmental Conference that aimed to address the Protection of Vegetables, Fruits, and Crops Against Bacteriological and Phytopathological Harm. This important conference was convened and took place in Montego Bay, Jamaica, in the year 1989, marking a significant step in global efforts to enhance pest management practices through cooperative international engagement and the promotion of biological control methods.

This initiative provided an invaluable platform for a diverse group of professionals, including scientists, plant protection engineers, pest management specialists, and economists, all hailing from various developing countries around the globe. These experts gathered to meet, collaborate, and exchange essential knowledge and extensive experience concerning Integrated Pest Management (IPM). During this significant event, numerous IPM projects were presented, showcased, and thoroughly discussed among the participants. Additionally, a small working group took the initiative to draft a comprehensive recommendation, which was later discussed and formally finalized for broader dissemination. The communication facilitated during this process occurred through both online channels managed by the FAO and through traditional paper-based correspondence. This latter avenue of communication ultimately culminated in the publication, in April 1991, of the proceedings titled "IPM in the Tropics: Current Status and Future Directions." The recommendation put forth to the FAO received unanimous acceptance from all involved parties, marking a pivotal moment that eventually led to the establishment of the FAO Inter-departmental group focused on 'Integrated Production and Protection', commonly referred to as IPPM (Integrated Production and Pest Management). Since its inception, FAO-IPPM has become increasingly active and has played a crucial role in promoting the principles and practices of IPM worldwide, operating from its headquarters situated in Rome. In its ongoing commitment to enhancing agricultural practices and pest management strategies, it has successfully established national IPM committees across various regions throughout the globe.

Conclusion

The convergence of human population growth, land use change, climate change, and the pathways of introduction and establishment of invasive species means that the burgeoning costs of invasive species will accelerate in the next few decades, especially in developing countries (Pretty & Pervez Bharucha, 2015). Future challenges of IPM implementation include insufficient coordination in integrative research, consultation and development; insufficient understanding of possible synergies and trade-offs of individual IPM measures; assessment and prediction of climate change impacts; insufficient consideration of individual and collective behaviours; and, insufficient explanation and communication of IPM to farmers.

Implicit in the concept of IPM is the need to consider all components of agroecosystems, including pests, crop and weed opponents, beneficial organisms, the physical environment and human management. IPM concentrates on the natural workings of ecosystems in preference to chemical treatments, while the additional consideration of the socio-economic and policy landscapes is a crucial oversight for the sustainability of the IPM paradigm (Damos et al., 2015). The primary goal of future crop protection systems is to optimise the balance between the benefits of pest control and the environmental costs of control. By combining results from multiple lines of reasoning, future systems should actively seek the resilience and sustainability outcomes of the IPM paradigm while protecting their formal structure.

Everything in agriculture is embedded in the social environment. Farmers have groups that share technology, support and advice, connections to marketing and extension agencies, and needs for acceptance in the wider society. Future crop protection systems need to build on these social networks, adapting structures to improve information sharing, technology management, and coherence throughout the production chain. Farm management is also a social operation, and the social networks that farmers presently have need to embed consideration of pests. Such strategies are a crucial prerequisite for farmers to become aware of and better manage pest populations. Formal cooperative group arrangements, governed by common goals and contracts, would need to be coupled with informal social structures. A simple way to structure and share collective expectations about pest management practices is through mapping farms according to key pests to individual groups, which will then proliferate in discussions about pest management.

Chapter 15
IPM Practical Guide for Farmers and Agricultural Engineers

Contents:

15.1. Step-by-Step Pest Management Planning

The following outlines a comprehensive step-by-step process for effective pest management planning. It is important to emphasize that there is no simple or universal formula that can create a pest management plan that will work in every context and for every situation. Nevertheless, this detailed outline can serve as a valuable springboard for your initial thoughts and actions. Additionally, it can be utilized to further develop and enhance one's skills in planning pest management strategies for present needs and future challenges. The first critical step in effective pest management planning is to clearly determine specific and measurable goals for your pest control and management activities. This initial step should be approached with careful consideration and should capture a comprehensive list of all desired outcomes for pest control, not merely those that may seem immediately reasonable or achievable. For instance, in the case of a potato grower aiming to control the Colorado potato beetle on the farm, the following goals may be included in the list: "Have pest populations consistently remain below levels that could result in crop loss," "decrease overall pesticide use significantly," and "integrate these efforts into an expanded Integrated Pest

Management (IPM) system." Embracing these ambitious goals will inevitably involve implementing some traditional pest control measures as well. The trick is to maintain focus on identifying and applying the most effective strategies for managing pest populations and ensuring that those carefully devised strategies are executed properly and efficiently.

The second step in the pest management process is to accurately determine the current level of pest pressure, which necessitates careful monitoring. This monitoring activity will help identify the current populations of pests and further analyze their nature, level of infestation, and spatio-temporal distribution. Such insights are invaluable as they enable timely and effective pest control actions to be implemented when necessary. At present, the most effective method of monitoring is primarily visual observation. It involves systematically recording the numbers of insects seen or encountered over a defined time period or at specific times and locations. This information is critical in assessing the current state of pest populations. In addition, several historical categories or groups of pesticides have been developed to target specific physiological sites in individual insects, making them particularly effective—these pesticides have been researched extensively and are well-understood. However, if the ongoing visual monitoring indicates that a particular group of pesticides is becoming less effective against the pest populations, then it is advisable to consider alternative options. This transition should be undertaken until the originally chosen pesticide becomes completely ineffective, ensuring that pest management efforts remain robust. The final step in the pest management planning process involves reflecting on the development and execution of the plan itself. It is essential to pinpoint which strategies seemed to be effective and to assess which aspects of the process could benefit from improvements in future iterations. New notes should be meticulously documented, capturing insights that can guide and enhance the planning and execution phases for next time, thus augmenting the overall effectiveness of pest management strategies.

15.2. Tools and Resources for Sustainable Practices

For sustainable practices to work effectively and yield fruitful results, farmers require more than merely the knowledge of the sustainable practices themselves. A robust and well-structured implementation process typically consists of a series of careful and critical steps: Initially, users are invited to attend planning meetings during which a comprehensive and meticulous assessment of baseline conditions is conducted. These planning meetings must be strategically designed to provide farmers with ample information about the proposed sustainable practices, ensuring they are well-informed regarding potential benefits, costs, and other important factors. As a consequence, the likely locations for successful implementation should be meticulously discovered and evaluated. During these planning meetings, significant support can be offered by alternative stakeholders to assist farmers in formulating realistic and achievable goals while also researching the preliminary design of the implementation activities. Farmers should be guided to identify all the necessary information, skills, and experience required for their initiatives, creating a detailed list that should be prepared for future negotiations among the involved stakeholders. By fostering collaboration and ensuring that farmers have access to the needed resources and expertise, it becomes easier to navigate the complexities of sustainable practices and facilitate their effective adoption in agricultural communities. (Quarcoo et al., 2017)

Once a support program is set up and put into action, farmers can be invited to engage in negotiation meetings with a variety of alternative stakeholders. These meetings are crucial as they provide a platform to discuss and outline the timetable for executing the implementation activities, alongside defining the specific roles and responsibilities of each stakeholder involved in the process. It is also essential that the farmers receive proper training not only on the sustainable practices they will be implementing but also on how to effectively negotiate with other stakeholders who are part of these discussions. In this context, drafts of all necessary agreements must be meticulously prepared by an independent facilitator. These drafts should then be reviewed collaboratively by all parties involved in the negotiations, allowing everyone to provide their input and necessary adjustments before moving to the finalized versions of the agreements, which will then be signed officially. Once the agreements have been signed by all parties, it is time for the implementation to commence. It is important during this phase to recognize that a great deal of patience and skill will be required to monitor the ongoing implementation process closely, ensuring that activities are closely aligned with the agreed-upon plans and objectives. Additionally, it will be necessary to adjust the implementation activities as required based on what is learned through ongoing observation and feedback. Entry into the monitoring process should ideally be postponed until all participating farmers are fully familiar and comfortable with living within the sustainable practices; otherwise, the monitoring efforts could easily derail and complicate the implementation process, leading to potential setbacks that could hinder progress. (Atcha-Ahowé et al., 2015).

Careful and thorough record keeping, alongside comprehensive analysis of the results, costs, successes, and failures during the implementation process, are absolutely essential for achieving more robust, effective, and impactful implementation of future sustainable practices. Establishing good personal relationships among all stakeholders helps to build vital trust and fosters the humility that is necessary to openly admit to mistakes and to proactively seek redress and support from one another. It is crucial that information regarding the implementation results and the various processes involved is disseminated as rapidly and widely as possible. This strategy maximizes the benefits emerging from the first successful implementation and encourages further engagement. When a more successful implementation is achieved, it can build significant momentum, contributing positively to ongoing efforts. However, it is important to understand that failures in implementation can lead to increased feelings of surprise and growing cynicism among stakeholders. Moreover, the full and active involvement of resource-poor farmers in every aspect of the implementation process is absolutely essential. Their involvement ensures that sustainable practices remain appropriate and relevant over time, thereby increasing their likelihood of being sustained. Additionally, it empowers farmers to take responsibility for the ongoing management of the practices that have been implemented, fostering a sense of ownership and commitment to their success.

15.3. Troubleshooting Common Pest Issues

Pests are defined as organisms that inflict injury or interfere with the growth and well-being of plants, animals, or humans. A broad range of examples of pest species can be found within this category, including various insects, fungi, nematodes, bacteria, different types of weeds, various bird species, and even some mammals. These pests cause significant and detrimental damage in a variety of ways, leading to serious consequences for agriculture and ecosystems alike. For instance, chewing insects possess the capability to remove leaf tissue, which ultimately results in defoliation and stress on the affected plants. On the other hand, sucking insects can extract sap, which weakens the plant substantially and may lead to growth deformities that impair its overall health and productivity. Fungi represent another major class of pests that can cause rots, wilts, and blights in plants, and some fungi can produce harmful toxins that render harvested produce unsafe and unfit for human consumption. Bacteria also play a role in pest-related damage, causing various afflictions such as wilts, rots, and cankers that adversely affect plant health. Additionally, nematodes are capable of feeding on the roots of plants, leading to symptoms such as yellowing, thinning, and stunting of growth. Competing with plants for essential resources, weeds pose another significant challenge as they strive for water, light, and nutrients. Not only do weeds compete directly, but they can also harbor diseases and insects that target and attack nearby plants, exacerbating the challenges that gardeners and farmers face in maintaining healthy crops and ecosystems.

Although damage inevitably occurs as a direct result of the activities carried out by various pests, it is notable that many pest species tend to remain at relatively low population levels. Economic thresholds are then carefully defined as the specific number of pests present within a designated area that will lead to damage occurring at a level that is equal to the costs involved in implementing various control measures. In other words, the economic threshold is defined as the particular pest numbers present when necessary management action must be taken to prevent future economic damage. For a significant number of pests, the mechanism employed for managing pest numbers relies heavily on damage/economic threshold analysis. Below these established threshold numbers, no specific actions are taken, and thus, nothing is done to control the pest populations. In contrast, when the pest numbers exceed the established threshold, appropriate control measures are promptly engaged and implemented. In this method of pest management, careful monitoring of plant damage is conducted, and when damage levels reach or exceed the specified threshold levels, necessary management action is taken decisively to mitigate further impact. (G. Higley, 1989).

This approach presents various problems and challenges. Significant pest damage can occur even while pest levels remain below established action thresholds. Complications arise particularly in multi-pest areas when the damage caused by one pest species is not adequately addressed, leading to the neglect of other species that could also be causing harm. The identification of damage symptoms serves as a monitoring tool, yet it comes with its own set of limitations. Effective monitoring necessitates a visual inspection of the crop, which inherently cannot differentiate between the monitoring of actual damage, the numbers of pests present, and the various management techniques that might be employed. Furthermore, colonization and infestation issues often emerge when monitoring pest populations, as actions may not be taken promptly—specifically, a delay of 3 to 5 days can occur because visible damage does not materialize immediately, resulting in the continued opportunity for pests to establish themselves further. (Saha et al., 2016).

15.4. Community Engagement and Education

Meeting the needs of diverse communities effectively requires a dedicated team of outreach specialists who possess a wide range of experiences, perspectives, and backgrounds. This team should also engage in ongoing monitoring and evaluation efforts, which include conducting pre- and post-event surveys to assess the impact and efficacy of their initiatives. Extension specialists play a crucial role in promoting and coordinating various community events while also taking responsibility for certifying trainers, managing event registrations, ensuring that adequate training materials are readily available for participants, and processing Continuing Education Unit (CEU) requests in an efficient and timely manner. To assist in these efforts, a comprehensive Community Integrated Pest Management (IPM) promotional brochure has been developed. This brochure serves as a compact yet informative overview of the Community IPM program, outlining its goals and benefits to the community. Additionally, a creative Video Contest named "Can you name that pest?" has been specifically designed to engage and educate youth about pest-related issues. The filming of these educational videos, which aim to demonstrate various pest problems, effective pest management strategies, the importance of natural enemies, the use of traps, and fundamental concepts in pest biology, is currently underway. Moreover, several other IPM outreach professionals have expressed their commitment to contributing valuable content by writing blogs as guest bloggers for social media platforms. These blogs will cover an array of new and diverse pest management topics, thereby broadening the scope of the program's outreach efforts and enhancing community engagement. These contributions will be seamlessly integrated into the overall outreach strategies of the program, ensuring that the community stays informed about the latest developments and best practices in integrated pest management. (M. Nowierski & J. Meyer, 2008)(Lampman et al., 2018).

Outdoor venues are also ideal sites for reaching a diverse range of audiences, and representatives of Community Integrated Pest Management (IPM) have participated in a number of booths at various festivals and large fairs. The Nubian HuB, which serves as an important community center established specifically to include and engage the often underserved immigrant community, has demonstrated considerable interest and enthusiasm in forging a collaborative partnership with the IPM program. Their goal is to educate its members about pest problems and effective pest management strategies. In collaboration with the dedicated volunteer effort that oversees the operation of the HuB, a comprehensive proposal for a series of informative collaborative presentations is currently being developed. To reinforce and build upon existing educational efforts, it is essential that IPM information be incorporated into programming initiatives that also address other crucial forms of pest management, such as ensuring compliance with local and state regulations to bolster a school trade program. Additionally, targeted agricultural IPM outreach is being diligently coordinated with suppliers who provide carefully chosen and effective pest management materials, ensuring the community has access to suitable resources for their pest control needs.

15.4.1. Farmer Workshops

Farmer participatory learning processes play a crucial role in making Integrated Pest Management (IPM) easy to understand and effectively apply at the farm level. In fact, the farmer field school (FFS) has emerged as a widely used experiential learning model for IPM within the context of West African agriculture. Within these schools, small groups, typically consisting of 10 to 30 farmers along with extension workers, come together to learn how to effectively integrate scientific information and advanced technologies with traditional farming practices. It is ideal that the training participants represent a diverse subset of farmers who they will frequently interact with within their communities. Participants should feel unrestrained and empowered to share the newly acquired knowledge and skills with others, and they should possess an openness to engaging in active group learning with a focus on problem-solving approaches. To facilitate this process, baseline surveys are carried out. These surveys aim to not only summarize the profiles of the proposed sites and the specific crops for training but also to provide a detailed overview of the current situation that needs to be improved. Additionally, these surveys are crucial in identifying potential participants along with their general characteristics, which include important demographic factors such as gender, age, education level, and farming experience, among others. The development of the curriculum is most effectively achieved when a collaborative team of three to eight individuals comprises scientists, trainers, extension agents, and representatives from the farming community work together. This collaborative team, along with the facilitators, will thoroughly analyze the data obtained from the baseline surveys. They will then identify the key constraints or issues that need addressing as well as the opportunities or solutions that can be incorporated into the curriculum to enhance learning outcomes. Once this collaborative team has drafted the curriculum, it is essential to share the resulting draft with FFS facilitators, extension staff, relevant representatives from the farming community, and other important stakeholders involved in the process. The curricula designed for FFS that target IPM cover a comprehensive range of aspects critical to successful farming practices. These aspects include vital elements such as land preparation techniques, soil health indicators, nursery management practices, strategies for crop growth, pest identification, and understanding natural enemy dynamics. Furthermore, it also emphasizes the integration of various options, whether they be separate or combined, and conducting a thorough economic analysis of IPM to evaluate its impact on farming outcomes. (Atcha-Ahowé et al., 2015).

IPM farmer study groups are typically quite small, usually consisting of around 15 to 38 farmers. These groups engage in three-in-one meetings, which consist of trainers, farmers, and extension staff coming together. During these collaborative meetings, the contents of representative schedules, which outline various aspects of the Integrated Pest Management (IPM) program, along with the progress and results of hands-on field trials, are collectively scrutinised by all participants. This collective approach fosters an environment of shared learning and continuous improvement. Through their own participatory research initiatives, farmers actively contribute to the efforts of researchers to develop and refine various IPM options that are suitable for their specific conditions and contexts. Sub-groups formed from the participants carry out dedicated field trials to rigorously test and validate these identified IPM options. Following topics suggested by farmers themselves—topics such as the dynamics of pest populations or the effectiveness and behaviours of trained farmer field trainers—control options related to pest management strategies are meticulously tested. Throughout this process, there is flexibility; if certain control options are deemed necessary or advantageous, they may be added to the trials, while others might be removed if they do not yield favorable results. The group processes that take place are critical to the success of these initiatives. They involve problem and issue identification, systematic data collection, and the integration of results with established scientific information. These discussions foster a vibrant dialogue around possible solutions and thoroughly examine the implications of various options, particularly focusing on their suitability as well as the benefit ratios of the trials conducted. It is not uncommon for other farmers from the same village to join these discussions, enriching the dialogue and sharing their experiences. During subsequent meetings, the contents of representative study plots are revisited and scrutinised even further. This allows participants to discuss the designs of the trials that have been conducted and strategise on future implementations, ensuring the continuous evolution of their practices and collaboration in managing pest-related challenges effectively.

15.4.2. Extension Services

IPM practices will be effectively promoted in West African vegetable production through a robust inter-disciplinary collaboration that intricately involves the fields of entomology, pathology, and agronomy. The practicalities of this multifaceted collaboration are greatly facilitated through the innovative development and comprehensive dissemination of the run-time IPM_vegetable software. This comprehensive guide is meticulously designed to elucidate how to thoroughly understand the pest status within a vegetable agroecosystem, enable the identification and assessment of the economic importance of various pests, and help in choosing and applying pertinent IPM options tailored against them. Through an engaging learning-by-doing process, the techniques and methodologies introduced will empower vegetable extension workers to markedly increase their knowledge and understanding of IPM principles and practices. Extension workers must realize that while they can certainly learn particular skills relevant to pest management, they cannot aspire to become specialist entomologists, pathologists, weed scientists, or agronomists in their own right. A heightened knowledge of the components that make up vegetable agroecosystems, as well as a deep understanding of the interactions that occur between these components, helps extension workers to regard the entire agroecosystem as a single integrated functional unit. To achieve this holistic perspective, they need to learn at least the foundational basics of the various approaches and hybridization of methods that are employed by their specialized colleagues in the field. Extension workers who are equipped with this holistic worldview can provide far better and more nuanced advice on the choice and application of IPM options in ways that will not only promote vegetable plant health but will also ensure the protection of human health and the broader environment. In order to furnish improved and reliable advice, they will need to make frequent and thorough farm visits, which are essential for gaining a solid understanding of the on-ground situation and enabling them to observe and assess the whole picture comprehensively. Enhanced results can only be achieved through steadfast cooperation with an array of diverse stakeholders. In many crucial respects, the challenges posed by vegetable pest problems in West Africa diverge remarkably from those encountered in industrialized nations or in regions characterized by a more temperate climate. The bio-physical conditions in West Africa differ substantially from those elsewhere, and the various forms of farming and production systems that are applied also showcase differing characteristics. Typically, in West Africa, fresh vegetables are produced primarily by smallholders who exhibit a high dependency on family labor. The markets for vegetables in this region are also characterized by significant differences that influence pest issues. These disparities carry profound effects on the pest problems encountered in the vegetables cultivated. Extension workers and farmers will learn through experience that a pest-free plot of vegetables does not inevitably translate into an increase in income; in fact, it may sometimes lead to a decrease in overall profits. The nature of vegetable pest problems is dynamic and ever-changing, with infestations that can alter dramatically over quite short time periods. New pests may emerge suddenly and proliferate quickly within the agroecosystem. Furthermore, inappropriate agronomic practices and the escalating impacts of climate change can elevate pests from minor status to an economically significant concern at an alarming rate. Therefore, frequent farm visits are essential to observe, note, and respond in a timely manner to these kinds of changes that can have substantial implications for production. Before imparting their knowledge to farmers, the extension workers must diligently monitor the vegetable fields of farmers and/or consult the crop inspection results that have been conducted by their colleagues in the field. They should pay particular attention to the pest problems observed in the fields and their changes over time. During their field monitoring visits, extension workers should also check for other potential problems that may occur, such as issues relating to pests, diseases, weeds, abiotic factors, varietal choice, and agronomic practices. It is crucial for extension workers to visit farmers on-site to establish a strong and productive working relationship. Thoughtful steps should be taken to optimize the selection of farmers for monitoring and/or intervention with IPM strategies. Pest lists hold significant importance in the context of international trade in vegetables. The field diagnostic methods and specimen collection techniques that are employed aim to significantly enhance the capacity of extension workers to accurately identify and assess the economic importance of pests. Extension workers should be encouraged to actively collate their observations into comprehensive pest lists and carefully curated reference collections of specimens. These pest lists serve as essential baseline resources, vital for informing governments and stakeholder groups about the economic risks associated with the international transfer of vegetable materials. Additionally, pest lists and reference collections constitute invaluable resources for extension workers to cross-check and advise on any observed changes in the incidence and diversity of pest populations. (Atcha-Ahowé et al., 2015)

References

Akuno, W., Onyango, C., & Obara, J., 2014. Influence of Selected Alternative Extension Approaches on the Acquisition of Knowledge, Skills for Agricultural Productivity in the Lake Victoria Region, Kenya. [PDF]

Alexandre Nogueira de Sá, L., Conceição Peres Young Pessoa, M., José de Moraes, G., Scardini Marinho-Prado, J., de Souza Prado, S., & Miriam de Vasconcelos, R., 2016. Estações quarentenárias e aspectos legais do uso de agentes de controle biológico no Brasil. [PDF]

Al-Kindi, K. M., Kwan, P., R. Andrew, N., & Welch, M., 2017. Impacts of human-related practices on Ommatissus lybicus infestations of date palm in Oman. ncbi.nlm.nih.gov

Ananth V. S., undefined & Vamsi D. K. K., undefined, 2021. Influence of quantity of additional food in achieving biological conservation and pest management in minimum-time for prey-predator systems involving Holling type III response. ncbi.nlm.nih.gov

Anderson, J. A., C. Ellsworth, P., C. Faria, J., P. Head, G., D. K. Owen, M., D. Pilcher, C., M. Shelton, A., & Meissle, M., 2020. Genetically engineered crops: importance of diversified integrated pest management for agricultural sustainability. [PDF]

Arakaki, A. & Nagamine, C., 2004. Calibrating Your Sprayer. [PDF]

Areola, F. O., 2010. Challenges to meeting best procedure in quarantine services delivery. [PDF]

Atcha-Ahowé, C., Baimey, H., Godonou, I., E. Goergen, G., James, B., Sikirou, R., & Toko, M., 2015. Integrated pest management in vegetable production: a guide for extension workers in West Africa. [PDF]

Atcha-Ahowé, C., Baimey, H., Godonou, I., E. Goergen, G., James, B., Sikirou, R., & Toko, M., 2015. Integrated pest management in vegetable production: a guide for extension workers in West Africa. [PDF]

Batz, P., Will, T., Thiel, S., Mark Ziesche, T., & Joachim, C., 2023. From identification to forecasting: the potential of image recognition and artificial intelligence for aphid pest monitoring. ncbi.nlm.nih.gov

Batz, P., Will, T., Thiel, S., Mark Ziesche, T., & Joachim, C., 2023. From identification to forecasting: the potential of image recognition and artificial intelligence for aphid pest monitoring. ncbi.nlm.nih.gov

Bradbury, S., 2014. Federal pesticide regulatory update: Moving to a national framework with local adaptations?. [PDF]

Campos, E. V. R., L. de Oliveira, J., Pascoli, M., de Lima, R., & F. Fraceto, L., 2016. Neem Oil and Crop Protection: From Now to the Future. ncbi.nlm.nih.gov

Candian, V., Giuseppe Pansa, M., Santoro, K., Spadaro, D., Briano, R., Peano, C., Tavella, L., & Tedeschi, R., 2021. First Multi-Target Application of Exclusion Net in Nectarine Orchards: Effectiveness against Pests and Impact on Beneficial Arthropods, Postharvest Rots and Fruit Quality. ncbi.nlm.nih.gov

Chandler, D., Bailey, A., (G. Mark) Tatchell, M., (Gill) Davidson, G., Greaves, J., & Grant, W., 2011. The development, regulation and use of biopesticides for integrated pest management. [PDF]

Chen, H., Weng, H., Zhu, H., Shen, S., & Li, W., 2024. Insights on droplet drift and effective utilization of pesticide in “Third Pole”: Qinghai-Tibet Plateau of China. ncbi.nlm.nih.gov

Cloyd, R. A., 2020. How Effective Is Conservation Biological Control in Regulating Insect Pest Populations in Organic Crop Production Systems?. ncbi.nlm.nih.gov

Coble, H. D., 1989. Weed Thresholds. [PDF]

Damos, P., Escudero Colomar, L. A., & Ioriatti, C., 2015. Integrated Fruit Production and Pest Management in Europe: The Apple Case Study and How Far We Are From the Original Concept?. ncbi.nlm.nih.gov

Demirozer, O., Tyler-Julian, K., Funderburk, J., Leppla, N., & Reitz, S., 2012. u3ciu3eFrankliniella occidentalisu3c/iu3e (Pergande) integrated pest management programs for fruiting vegetables in Florida. [PDF]

Doran, J. W. & R. Zeiss, M., 2000. Soil health and sustainability: managing the biotic component of soil quality. [PDF]

Dumont, Y. & F. Oliva, C., 2024. On the impact of re-mating and residual fertility on the Sterile Insect Technique efficacy: Case study with the medfly, Ceratitis capitata. ncbi.nlm.nih.gov

Dunn, A., 2017. Opportunities and needs for biological control in New York. [PDF]

Fernández-Grandon, G.M., J. Harte, S., Ewany, J., Bray, D., & C. Stevenson, P., 2020. Additive Effect of Botanical Insecticide and Entomopathogenic Fungi on Pest Mortality and the Behavioral Response of Its Natural Enemy. ncbi.nlm.nih.gov

Francis, F., Jacquemyn, H., Delvigne, F., & Lievens, B., 2020. From Diverse Origins to Specific Targets: Role of Microorganisms in Indirect Pest Biological Control. ncbi.nlm.nih.gov

Frank, D. L., Starcher, S., & S. Chandran, R., 2020. Comparison of Mating Disruption and Insecticide Application for Control of Peachtree Borer and Lesser Peachtree Borer (Lepidoptera: Sesiidae) in Peach. ncbi.nlm.nih.gov

Gold, R. E., 1981. IPM - An Overview. [PDF]

Grasswitz, T. R., 2019. Integrated Pest Management (IPM) for Small-Scale Farms in Developed Economies: Challenges and Opportunities. ncbi.nlm.nih.gov

Grégoire, J. C., Caiti, E., Hasbroucq, S., Molenberg, J. M., & Willenz, S., 2022. When the Beetles Hit the Fan: The Fan-Trap, an Inexpensive, Light and

Scalable Insect Trap under a Creative Commons License, for Monitoring and Experimental Use. ncbi.nlm.nih.gov

Grisso, R., Hay, D. L., J. Jasa, P., K. Koelsch, R., Skipton, S., & Woldt, W., 1998. EC98-767 Farm*A*Syst Nebraska's System for Assessing Water Contamination Fact Sheet 4: Improving Pesticide Storage and Handling. [PDF]

Hamlyn, O., 2015. Sustainability and the failure of ambition in European pesticides regulation. [PDF]

Harvey-Samuel, T., Ant, T., & Alphey, L., 2017. Towards the genetic control of invasive species. ncbi.nlm.nih.gov

Hassan Mirza, J., Kamran, M., & Jaber Alatawi, F., 2021. Phenology and abundance of date palm mite Oligonychus afrasiaticus (McGregor) (Acari: Tetranychidae) in Riyadh, Saudi Arabia. ncbi.nlm.nih.gov

Higley, L. G., 1989. Making Management Decisions for Pests: Past, Present, and Future Approaches. [PDF]

Husain, M., G. Rasool, K., D. Sutanto, K., O. Omer, A., Tufail, M., & S. Aldawood, A., 2024. Laboratory Evaluation of Indigenous and Commercial Entomopathogenic Nematodes against Red Palm Weevil, Rhynchophorus ferrugineus (Coleoptera: Curculionidae). ncbi.nlm.nih.gov

Jeger, M., Bragard, C., Caffier, D., Candresse, T., Chatzivassiliou, E., Dehnen-Schmutz, K., Grégoire, J. C., Anton Jaques Miret, J., MacLeod, A., Navajas Navarro, M., Niere, B., Parnell, S., Potting, R., Rafoss, T., Rossi, V., Urek, G., Van Bruggen, A., Van Der Werf, W., West, J., Winter, S., Hart, A., Schans, J., Schrader, G., Suffert, M., Kertész, V., Kozelska, S., Rosaria Mannino, M., Mosbach-Schulz, O., Pautasso, M., Stancanelli, G., Tramontini, S., Vos, S., & Gilioli, G., 2018. Guidance on quantitative pest risk assessment. ncbi.nlm.nih.gov

Kim, B. J., Cheon Hong, S., Egger, D., S. Katsar, C., & L. Griffin, R., 2019. Predictive Modeling and Categorizing Likelihoods of Quarantine Pest

Introduction of Imported Propagative Commodities from Different Countries. ncbi.nlm.nih.gov

Kuesel, R., Scott Hicks, D., Archer, K., Sciligo, A., Bessin, R., & Gonthier, D., 2019. Effects of Fine-Mesh Exclusion Netting on Pests of Blackberry. ncbi.nlm.nih.gov

Lampman, J., Braband, L., Dunn, A., Frye, M., Gangloff-Kaufmann, J., & Grant, J., 2018. Community IPM Outreach Efforts for 2018. [PDF]

López, O., Onofre Martinez Rach, M., Migallon ., H., Jose Perez Malumbres, M., Miguel Bonastre Pina, A., & José Serrano Martín, J., 2012. Monitoring pest insect traps by means of low-power image sensor technologies. [PDF]

López-Castillo, L.M., E Silva-Fernández, S., Winkler, R., Bergvinson, D., T Arnason, J., & García-Lara, S., 2018. Postharvest insect resistance in maize. [PDF]

Mariyono, J., C Luther, G., Bhattarai, M., Ferizal, M., Jaya, R., & Fitriana, N., 2013. Farmer Field Schools on Chili Peppers in Aceh, Indonesia: Activities and Impacts. [PDF]

Matten, S., P. Head, G., & D. Quemada, H., 2008. How Governmental Regulation Can Help or Hinder the Integration of Bt Crops within IPM Programs. ncbi.nlm.nih.gov

Matten, S. M., L. Hellmich, R., & Reynolds, A., 2004. Current resistance management strategies for Bt corn in the United States. [PDF]

Matten, S. R., P. Head, G., & D. Quemada, H., 2008. How Governmental Regulation Can Help or Hinder the Integration of Bt Crops within IPM Programs. ncbi.nlm.nih.gov

Mitchell, H. J. & Bartsch, D., 2020. Regulation of GM Organisms for Invasive Species Control. ncbi.nlm.nih.gov

Morrison, W. R., Bruce, A., V. Wilkins, R., E. Albin, C., & H. Arthur, F., 2019. Sanitation Improves Stored Product Insect Pest Management. ncbi.nlm.nih.gov

Muhibuddin, A.A., Addina, L., L. Abadi, A., & Ahmad, A., 2011. Biodiversity of Soil Fungi on Integrated Pest Management Farming System. [PDF]

Na, S. M., Im, G. I., Lee, W. S., & Kim, D. G., 2024. Assessment of Attractant Combinations for the Management of Red Palm Weevils (Rhynchophorus ferrugineus) in the United Arab Emirates. ncbi.nlm.nih.gov

Nowierski, R. M. & J. Meyer, H., 2008. Establishing Inter-agency, Multidisciplinary Areawide Pest Management Programmes. [PDF]

Nurashikin-Khairuddin, W., Noor Aishikin Abdul-Hamid, S., Saiful Mansor, M., Bharudin, I., Othman, Z., & Jalinas, J., 2022. A Review of Entomopathogenic Nematodes as a Biological Control Agent for Red Palm Weevil, Rhynchophorus ferrugineus (Coleoptera: Curculionidae). ncbi.nlm.nih.gov

Okigbo, B., 2018. Multidisciplinary approach to pest management and the African farmer. [PDF]

Parsa, S., Morse, S., Bonifacio, A., C B Chancellor, T., Condori, B., Crespo-Pérez, V., L A Hobbs, S., Kroschel, J., N Ba, M., Rebaudo, F., G Sherwood, S., J Vanek, S., Faye, E., A Herrera, M., & Dangles, O., 2014. Obstacles to integrated pest management adoption in developing countries. [PDF]

Payne, B., 2019. Almond's production in the California Central Valley. [PDF]

Pope, R. & DeWitt, J., 2005. ISU Pest Management Monitoring Programs in Iowa: Where and How to Get This Information for 2006. [PDF]

Pretty, J. & Pervez Bharucha, Z., 2015. Integrated Pest Management for Sustainable Intensification of Agriculture in Asia and Africa. ncbi.nlm.nih.gov

Quarcoo, F., Bonsi, C., Nii O Tackie, D., A Hill, W., Wall, G., & Hunter, G., 2017. Economies of Scale in Integrated Pest Management in Vegetable and Fruit Production. [PDF]

Quinto Cánovas, J., María Pineda Gómez, A., & Ángeles Marcos-García, M., 2010. Control natural de plagas en cultivos mediterráneos. [PDF]

Rider, A. R. & C. Dickey, E., 1982. Field Evaluation of Calibration Accuracy for Pesticide Application Equipment. [PDF]

Robotham, M., Evensen, C., & J. Cox, L., 2000. Yard and Garden Pest Management. [PDF]

Rourke, K., 2017. ATTRACTING BENEFICIAL INSECTS TO YOUR FARM: A COMPARISON OF HABITAT MODIFICATION STRATEGIES. [PDF]

Russell, D.A., R Kranthi, K., Surulivelu, T., R Jadhav, D., Regupathy, A., & Singh, J., 2000. Developing and implementing insecticide resistance management practices in cotton ICM programmes in India. [PDF]

Saha, S., Pfeufer, E., Bessin, R., Wright, S., & A. Kaiser, C., 2016. An IPM Scouting Guide for Common Problems of High Tunnel and Greenhouse Vegetable Crops in Kentucky. [PDF]

Schuldt, A., Ebeling, A., Kunz, M., Staab, M., Guimarães-Steinicke, C., Bachmann, D., Buchmann, N., Durka, W., Fichtner, A., Fornoff, F., Härdtle, W., R. Hertzog, L., Klein, A. M., Roscher, C., Schaller, J., von Oheimb, G., Weigelt, A., Weisser, W., Wirth, C., Zhang, J., Bruelheide, H., & Eisenhauer, N., 2019. Multiple plant diversity components drive consumer communities across ecosystems. ncbi.nlm.nih.gov

Sharma, H. C., 2013. Climate Change Effects on Insects: Implications for Crop Protection and Food Security. [PDF]

Soares dos Anjos Marques, A., Lopes-da-Silva, M., Gonzaga, V., Rausch Fernandes, F., Polo Benito, N., & Ferraz de Arruda Veiga, R., 2016. Biological fundamentals, operational tools, and innovation in plant quarantine. [PDF]

Speedy, C., Day, T., & Innes, J., 2007. Pest Eradication Technology – The Critical Partner to Pest Exclusion Technology: The Maungatautari Experience. [PDF]

Stetkiewicz, S., Bruce, A., J Burnett, F., A Ennos, R., & F E Topp, C., 2018. Perception vs practice: Farmer attitudes towards and uptake of IPM in Scottish spring barley. [PDF]

Suzuki, T., 2012. Environmental Engineering Approaches toward Sustainable Management of Spider Mites. ncbi.nlm.nih.gov

Takeuchi, Y., Tripodi, A., & Montgomery, K., 2023. SAFARIS: a spatial analytic framework for pest forecast systems. ncbi.nlm.nih.gov

Talwana, H., Masiko, M., Dramani, S., & Edimu, F., 2018. Abundance and diversity of arthropod pests infesting stored maize in smallholder farmers and traders systems highlight critical points for pest management in Uganda: Presentation. [PDF]

Therese Langeland Næss, E., 2016. Molecular analysis of predation by anthocorid bugs on the pear psyllid Cacopsylla pyri (Homoptera, Psyllidae).. [PDF]

Thorburn, C., 2015. The Rise and Demise of Integrated Pest Management in Rice in Indonesia. ncbi.nlm.nih.gov

Tonnang, H. E. Z., D B Hervé, B., Biber-Freudenberger, L., Salifu, D., Subramanian, S., B Ngowi, V., Y A Guimapi, R., Anani, B., M M Kakmeni, F., D Affognon, H., Niassy, S., Landmann, T., T Ndjomatchoua, F., A Pedro, S., Johansson, T., M Tanga, C., Nana, P., M Fiaboe, K., F Mohamed, S., K Maniania, N., V Nedorezov, L., Ekesi, S., & Borgemeister, C., 2017. Advances in crop insect modelling methods—Towards a whole system approach. [PDF]

Umina, P. A., Jenkins, S., McColl, S., Arthur, A., & A. Hoffmann, A., 2015. A Framework for Identifying Selective Chemical Applications for IPM in Dryland Agriculture. ncbi.nlm.nih.gov

Vimala Devi, P. S., V Ranga Rao, G., Gopalakrishnan, S., & Sivakumar, G., 2012. Environmental Impact of Microbial Pesticides. [PDF]

White, G. B. & S. Lazarus, S., 1986. TB116: Integrated Systems for Managing Potatoes in the Northeast. [PDF]

Wolf, R. E., 1996. New Designs in Pesticide Application Equipment for More Efficient Control and Reduced Drift. [PDF]

Wright, R. J., D. Danielson, S., & F. Witkowski, J., 1991. EC91-1509 1991 Insect Management Guide for Nebraska : Corn and Sorghum. [PDF]

Wright, R.J., 1995. EC95-1560 Use of Cultural Practices in Crop Insect Pest Management. [PDF]

Wright, R.J., 1995. G95-1251 Biological Control of Insect and Mite Pests. [PDF]

York State Integrated Pest Management Program, N., 2016. Think IPM. [PDF]

Zhang, H., 1970. Promoting integrated pest management in arable fields. [PDF]

Ziska, L. H., A. Bradley, B., D. Wallace, R., T. Bargeron, C., H. LaForest, J., A. Choudhury, R., A. Garrett, K., & E. Vega, F., 2018. Climate Change, Carbon Dioxide, and Pest Biology, Managing the Future: Coffee as a Case Study. [PDF]

Additional References

Al Basir, F., Chowdhury, J., & Torres, D. F. M. (2023). Dynamics of a Double-Impulsive Control Model of Integrated Pest Management Using Perturbation Methods and Floquet Theory. PDF Link

Alemu, K. (2015). System Management Approach in Crop Production and Protection. Link to Paper

Alexandre Nogueira de Sá, L., Conceição Peres Young Pessoa, M., José de Moraes, G., Scardini Marinho-Prado, J., de Souza Prado, S., & Miriam de Vasconcelos, R., 2016. Estações quarentenárias e aspectos legais do uso de agentes de controle biológico no Brasil. [PDF]

Allen-Perkins, A. & Estrada, E. (2019). Mathematical modeling for sustainable aphid control in agriculture via intercropping. Link to Paper

Amek, T. (2011). a case of small scaler rural farmers in Kenya. Read More

Anderson, J. A., Ellsworth, P. C., Faria, J. C., Head, G. P., Owen, M. D. K., Pilcher, C. D., Shelton, A. M., & Meissle, M. (2020). Genetically engineered crops: importance of diversified integrated pest management for agricultural sustainability. PDF Link

Areola, F. O., 2010. Challenges to meeting best procedure in quarantine services delivery. [PDF]

Atcha-Ahowé, C., Baimey, H., Godonou, I., Goergen, G. E., James, B., Sikirou, R., & Toko, M. (2015). Integrated pest management in vegetable production: a guide for extension workers in West Africa. PDF Link

Bailey, A., Bertaglia, M., Fraser, I., Sharma, A., & Douarin, E. (2009). Integrated Pest Management Portfolios in UK Arable Farming: Results of a Farmer Survey. PDF Link

Baumgärtner, J., Schulthess, F., & Xia, Y. (2017). Integrated Arthropod Pest Management Systems for Human Health Improvement in Africa. PDF Link

Bradbury, S.P., 2014. Federal pesticide regulatory update: Moving to a national framework with local adaptations?. [PDF]

Brewer, G. J. (2005). Integrated Pest Management in the Global Arena. Read more here

Campbell, J. B., & McNeal, C. D. (1980). EC80-1536 A Guide to Integrated Pest Management at Feedlots and Dairies. Read more here

Centre for Agricultural and Rural Cooperation. (2015). Q&A: The status of ICTs in integrated pest management. Link to Source

Damos, P., Escudero Colomar, L. A., & Ioriatti, C. (2015). Integrated Fruit Production and Pest Management in Europe: The Apple Case Study and How Far We Are From the Original Concept?. Link

Dijkxhoorn, Y., Bremmer, J., & Kerklaan, E. (2013). Towards Integrated Pest Management in East Africa: a feasibility study. Link

Gold, R. E. (1981). IPM - An Overview. PDF Link

Gold, R. E., 1981. IPM - An Overview. [PDF]

Grasswitz, T.R. (2019). Integrated Pest Management (IPM) for Small-Scale Farms in Developed Economies: Challenges and Opportunities. Read More

Higley, L.G. 1989. Making Management Decisions for Pests: Past, Present, and Future Approaches. Link to the article

Jeger, M., Bragard, C., Caffier, D., Candresse, T., Chatzivassiliou, E., Dehnen-Schmutz, K., Grégoire, J. C., Anton Jaques Miret, J., MacLeod, A., Navajas Navarro, M., Niere, B., Parnell, S., Potting, R., Rafoss, T., Rossi, V., Urek, G., Van Bruggen, A., Van Der Werf, W., West, J., Winter, S., Hart, A., Schans, J., Schrader, G., Suffert, M., Kertész, V., Kozelska, S., Rosaria Mannino, M., Mosbach-Schulz, O., Pautasso, M., Stancanelli, G.,

Tramontini, S., Vos, S., & Gilioli, G., 2018. Guidance on quantitative pest risk assessment. ncbi.nlm.nih.gov

Kim, B. J., Cheon Hong, S., Egger, D., S. Katsar, C., & L. Griffin, R., 2019. Predictive Modeling and Categorizing Likelihoods of Quarantine Pest Introduction of Imported Propagative Commodities from Different Countries. ncbi.nlm.nih.gov

Matten, S. M., Hellmich, R. L., & Reynolds, A. (2004). Current resistance management strategies for Bt corn in the United States. Link to Article

Matten, S. R., P. Head, G., & D. Quemada, H., 2008. How Governmental Regulation Can Help or Hinder the Integration of Bt Crops within IPM Programs. ncbi.nlm.nih.gov

Nowierski, R. M. & J. Meyer, H., 2008. Establishing Inter-agency, Multidisciplinary Areawide Pest Management Programmes. [PDF]

Nowierski, R. M. & Meyer, H. J. (2008). Establishing Inter-agency, Multidisciplinary Areawide Pest Management Programmes. Link to Paper

Okigbo, B. (2018). Multidisciplinary approach to pest management and the African farmer. Link to Paper

Parsa, S., Morse, S., Bonifacio, A., Chancellor, T. C. B., Condori, B., Crespo-Pérez, V., Hobbs, S. L. A., Kroschel, J., Ba, M. N., Rebaudo, F., Sherwood, S. G., Vanek, S. J., Faye, E., Herrera, M. A., & Dangles, O. (2014). Obstacles to integrated pest management adoption in developing countries. PDF Link

Pretty, J. & Pervez Bharucha, Z. (2015). Integrated Pest Management for Sustainable Intensification of Agriculture in Asia and Africa. Link to Paper

Quarcoo, F., Bonsi, C., Nii O Tackie, D., A Hill, W., Wall, G., & Hunter, G. (2017). Economies of Scale in Integrated Pest Management in Vegetable and Fruit Production. Read More

Ramsden, M., Furiosi, M., Debenedettis, P., Stojacic, I., Mendes, M., Munier-Jolain, N., & Caffi, T. (2022). Development of an online pan-European Integrated Pest Management Resource Toolbox. Read more here

Razzaghi Borkhani, F., Rezvanfar, A., Shabanali Fami, H., & Pouratashi, M. (2013). Social Factors Influencing Adoption of Integrated Pest Management (IPM) Technologies by Paddy Farmers. Read More

Rebaudo, F. & Dangles, O. (2011). Coupled Information Diffusion–Pest Dynamics Models Predict Delayed Benefits of Farmer Cooperation in Pest Management Programs. Read More

Riggi, L. (2017). Integrated pest management across spatial scales. Link to Paper

Soares dos Anjos Marques, A., Lopes-da-Silva, M., Gonzaga, V., Rausch Fernandes, F., Polo Benito, N., & Ferraz de Arruda Veiga, R., 2016. Biological fundamentals, operational tools, and innovation in plant quarantine. [PDF]

Speedy, C., Day, T., & Innes, J., 2007. Pest Eradication Technology – The Critical Partner to Pest Exclusion Technology: The Maungatautari Experience. [PDF]

Stetkiewicz, S., Bruce, A., J Burnett, F., A Ennos, R., & F E Topp, C. (2018). Perception vs practice: Farmer attitudes towards and uptake of IPM in Scottish spring barley. Link to Paper

Umina, P. A., Jenkins, S., McColl, S., Arthur, A., & A. Hoffmann, A., 2015. A Framework for Identifying Selective Chemical Applications for IPM in Dryland Agriculture. ncbi.nlm.nih.gov

Wise, K., Waldron, K., & Nedrow, J. (2018). Elements of IPM for Field Corn in New York State. Link to Article

York State Integrated Pest Management Program, N. (2016). Think IPM. Link to Article

Zhang, H. (1970). Promoting integrated pest management in arable fields. Link

FSC
www.fsc.org
MIX
Papier | Fördert gute Waldnutzung
FSC® C083411

Kolibri 360 GmbH
Ferdinand-Jühlke-Straße 7
99095 Erfurt, Deutschland
produktsicherheit@kolibri360.de

Druck:
CPI Druckdienstleistungen GmbH
im Auftrag der
Kolibri 360 GmbH
Ferdinand-Jühlke-Str. 7
99095 Erfurt